DRYING FOOD FOR PROFIT

This book is dedicated to the memory of
Dr Mario Molina of the Institute of Nutrition of
Central America and Panama in Guatemala. Mario was
a respected scientist and teacher, and a good close
friend. We worked for many years together on many
issues, including small-scale drying problems, and
shared both successes and failures. With Mario
this was always done with a smile and a joke.

DRYING FOOD FOR PROFIT

A Guide for Small Businesses

Barrie Axtell

with contributions from
Andrew Russell

**Practical
ACTION
PUBLISHING**

Practical Action Publishing Ltd
25 Albert Street, Rugby, CV21 2SD, Warwickshire, UK
www.practicalactionpublishing.com

First published 2002

ISBN 13 Paperback: 9781853395208
ISBN Library Ebook: 9781780444819
Book DOI: http://dx.doi.org/10.3362/9781780444819

Since 1974, Practical Action Publishing has published and disseminated books and information in support of international development work throughout the world. Practical Action Publishing is a trading name of Practical Action Publishing Ltd (Company Reg. No. 1159018), the wholly owned publishing company of Practical Action. Practical Action Publishing trades only in support of its parent charity objectives and any profits are covenanted back to Practical Action (Charity Reg. No. 247257, Group VAT Registration No. 880 9924 76).

Typeset by Dorwyn Ltd., Rowlands Castle, Hants

Contents

Acknowledgements

We wish to thank the Commonwealth Science Council, the Intermediate Technology Development Group and the National Lotteries Charities Board (NLCB) for financial assistance in the preparation and production of this book. Thanks are also due to Peter de Groot for the Uganda case study, G.R. Whitby for the use of market share tables, Mr Otim for data on the Ugandan biomass dryer, Oliver Hensel for the information on chimney draughts and Matthew Whitton for his illustrations. Thanks are also due to Sue Azam Ali, Peter Fellows, Peter de Groot, Oliver Hensel, Kate Potts and David Trim for commenting on chapters of the book. Grateful thanks must also go to my wife, Ruth, for all her patience and support.

Finally, we would like to express our thanks to many contacts around the world for opening our eyes to the real problems of small-scale entrepreneurs. Without their experiences this book could never have been written.

The psychrometric charts (Figures 2.2, 7.1, 7.2 and 7.3) are reproduced by permission of the Chartered Institute of Building Services Engineers. Pads of charts for calculation and record purposes are available from CIBSE, 222 Balham High Road, London SW12 9BS, UK.

Crown copyright material is reproduced with the permission of the Controller of Her Majesty's Stationery Office.

1 Introduction

1.1 The aim and scope of this book

IT IS HOPED that this publication will be helpful to the owners of enterprises involved in, or considering establishing, commercial facilities to dry foods to generate income, as well as agencies and institutions involved in promoting small-scale food processing for job creation. Its focus is on small-scale sustainable technologies that build on indigenous skills with minimum reliance on external inputs. In addition to examining improvements that can be made to traditional technologies, it stresses the importance of controllable, hygienic drying systems that can meet the demands of consumers and buyers who are increasingly conscious of food quality and safety.

It is essential, when considering processing any food for income, not only to examine and understand the technical aspects of that processing, but also aspects such as market studies, marketing, packaging, quality assurance and business economics must be addressed. Business success or failure will, in most cases, depend on these factors rather than upon the technology used. This publication thus briefly examines all of these areas rather than concentrating solely on the technology of drying. Numerous reference texts are available that cover the technology in more depth, should the reader require further information.

In most countries food dryers can be either completely or largely constructed locally. This requires a close working relationship between the processor and the engineers, who need to understand how to design equipment to meet a required demand. A chapter dedicated to the design of small mechanical drying equipment is included to provide guidance to engineers. As yet, in many countries, locally built dryers are not available as standard items and engineering workshops tend to construct to order. In such situations processors will need to be sure that the workshop understands all their requirements.

Engineers require a basic knowledge of the relevant food technology in order to incorporate good hygienic design into the equipment they make, and the owners of enterprises must understand the language and constraints of the engineer. By working in partnership, drying equipment can be locally designed, constructed and, most importantly, maintained. The final chapter of this publication provides guidelines to engineers and processors for the local design and construction of mechanical dryers. Considerable effort has been made to simplify the complex mathematics that is found in standard works on dehydration. The reader who can add, subtract, divide, multiply and calculate percentages will be able to carry out all the calculations needed to design a dryer for a given application.

It is also hoped that this book will prove useful to food research institutions and food technology students.

1.2 Traditional food drying

The preservation of foods by drying is without doubt the oldest method practised and it remains the most commonly used method worldwide. While grains and pulses are the most important, in terms of tonnage, the range of products dried is wide and includes meats, fish, fruits, vegetables, spices, nuts and milk.

Traditional dried foods and the methods used to produce them are related to the local climate; using the heat of the sun, shade, low humidity and natural draught. Figure 1.1 shows a traditional method of shade drying

Figure 1.1 Traditional shade drying of corn (ITDG–Paul Harris).

corn. In some cases the heat of cooking fires, possibly taking advantage of the preservative nature of smoke, is used. As would be expected, traditional dry food products are mainly produced in arid areas with abundant sunshine. Historical evidence exists of practices that have been carried out for thousands of years and still remain in use today. In Peru, for example, a dried potato product, *papa seca*, has been found in ancient Inca graves in such good condition that it could be rehydrated in water after one thousand years. *Papa seca* remains a staple element in the diet of people in the high Andean Mountains. Another interesting Andean product is *chuno*, which is made from potatoes that grow at altitudes of 4000 m or more. *Chuno* is made by a process, very similar to modern freeze drying, in which the potatoes are laid out at night in sub-zero temperatures to freeze. The next day the combination of the heat of the sun and reduced pressure, due to the altitude, allows the ice to evaporate directly to vapour without passing through the water phase.

In northern Kenya the nomadic Maasai people who travel through their arid pastures prepare dried milk and blood to produce lightweight, highly nutritious foods ideally suited to their way of life.

Traditional dried foods are less common in humid areas of the world. Cassava, for example, is processed and dried in West Africa and Brazil by toasting over a fire. In many northern countries meat and fish are preserved by a combination of salting, smoking and drying for use in the hungry winter months.

It is important to understand that all traditional dry foods have been developed in a way that takes into account the local climatic conditions. There can be a very high risk of food poisoning if the production of foods, such as those made by the Maasai in a dry climate, were to be transferred to a humid area without the use of modern storage and packaging technologies.

1.3 Need for improvements and opportunities

Modern production, distribution, packaging and storage technologies have led to the development of very large companies, often supplying global markets. To meet consumer demands, new drying technologies have been developed including roller dryers, spray dryers and freeze dryers. In industrialized countries, in practically all cases, the technologies used are large scale and highly automated.

Social change and urbanization occurring worldwide is resulting in consumers requiring new types of food, for example those which are quick and convenient to prepare. Opportunities may exist for these demands to be met through local manufacture, rather than by expensive imports.

The increasing use of dried products, such as fruits and vegetables from tropical countries, in a wide range of foods including breakfast cereals, dry soup mixes and health foods could open market opportunities for developing countries, provided that the required quality standards can be met by the producer. It is possible that, in time, the increasing demand by consumers for organically produced foods may also offer opportunities, as in many developing countries crops are traditionally grown without the use of herbicides and pesticides.

The production of dehydrated foods has several advantages for small and medium-scale producers.

o The technology is relatively simple and readily understood.
o The equipment required is of moderate cost and, in most countries, can be largely constructed locally.
o In general, the final products carry a low public health risk.
o Packaging costs can be low if plastic bags are used.
o The final product weight is low, thus reducing transport costs.
o Quality control checks are relatively simple.

The main disadvantages of dried foods relate to their market.

o Local markets may be limited, for example the sale of dry fruits as snacks or as an ingredient for bakeries.
o Some foods, such as dehydrated potato, carrot and onion are produced using intensive agriculture and are dried on a very large scale. The economies of scale mean that small producers may be unable to compete with such internationally traded bulk products.

1.4 Preservation and value addition by drying

Foods preserved by drying can be divided into two major groups. In the first, and by far the largest group, the objective of drying is **solely to preserve the food** by reducing the moisture content to a level that will prevent deterioration. Examples include rice, wheat, beans and maize. In many parts of the world these crops are dried by simply laying them out on mats, concrete floors or even roads. Little or no value is added by drying, the objective is simply to prevent post-harvest loss. In many countries there is no price differential in markets between, for example, well-dried and poorly dried rice or corn.

There is a commonly held opinion that small farmers or farmer groups need appropriate dryers for basic crops. Farmers often express keen interest, particularly in or after a season when bad weather has resulted in losses. However, in reality they are, in most cases, reluctant to invest in equipment to dry substantial quantities of basic crops or are unable to meet the fuel costs. As little or no value will be added to the product, the drying equipment will only be used when farmers face total or partial loss. Artificial drying of staple crops for preservation and food security is currently only generally successful on a very large scale in, for example, government or marketing board silos, where the economies of scale apply.

In the second group of foods, drying is primarily for income or to preserve the food for use in the home or community when seasonality causes prices to rise. Typical products include fruit, vegetables, nuts, spices, fish, herbs and medicinal plants. In these cases **value is added by processing** and this added value can be sufficient to cover processing costs and to provide income, or savings, for the owner of the enterprise.

In the next chapter the basic technical aspects of food drying are examined. Subsequent chapters explain how to carry out market surveys, and hence estimate the expected scale of production required. Typical small-scale drying technologies are then examined, followed by examples of processing common foods together with issues, such as quality, packaging and product costing and the preparation of a business plan. The final chapter provides guidelines to those wishing to design and construct small mechanical dryers.

2 Basic principles of food drying

THE WAY THAT different types of foods dry is controlled by four main factors:

o the physical structure and chemical composition of the food being dried
o the temperature, humidity and quantity of air used for drying
o the size of the particles of food being dried
o factors related to the geometry, and hence airflow patterns, in the dryer used.

This chapter examines each of these areas together with other general technical areas related to drying.

2.1 Food composition

All foods consist of components such as proteins, carbohydrates, fats, moisture, vitamins and minerals. As indicated in Table 2.1, the ratios of these components within foods vary considerably. Some of these components have a major influence on the way that moisture can be released from a food as it dries.

Table 2.1 Typical chemical composition of some common foods

Food	Protein (%)	Fat (%)	Fibre (%)	Total carbohydrate (%)	Moisture (%)
Cabbage	1.6	0.4	1.0	5.6	91.0
Avocado	1.3	8.2	1.3	5.7	84.0
Red pepper	2.8	1.7	4.0	9.9	85.0
Onion	1.4	0.2	0.8	9.7	88.0
Banana	1.1	0.3	3.4	19.2	70.7
Cassava	1.0	0.2	–	31.4	65.5
Potato	2.1	0.1	2.1	20.8	75.8
Mushroom	1.8	0.6	2.5	0	91.5
Oily fish	21.3	8.2	–	–	69.4
Cod	17.5	0.3	–	–	81.3

Source: McCance and Widdowson 1991

6

Carbohydrates are present in all living matter and can be divided into three main types:

○ *Simple sugars*, such as sucrose which is found in sugar cane and sweet fruits. Sugars have the chemical ability to bind to water, thus making it difficult to remove moisture by drying. In addition, sugars that are soluble in water tend to move towards the surface of a particle of food as it dries, forming a layer that binds the water and so slows the drying rate. This phenomenon, which is known as case hardening, is discussed more fully in Section 2.5.

○ *Starches* are long chains of simple sugars that predominate in root crops such as potato, cassava and yams. Green bananas are unusual fruits in that they contain high levels of starch. Starchy foods dry considerably more quickly than sweet fruits.

○ *Fibre* and *cellulose* are also made up of long chains of sugar molecules and predominate in leafy materials and, to a lesser extent, root crops. The open fibrous structure of foods such as onion or celery, for example, permits water to move easily to the surface where it can evaporate. Fibrous herbs and leafy foods are, in addition, extremely thin and thus dry very rapidly. Some leaves, however, have a waxy surface coating which does not allow the passage of moisture. In such cases drying is very slow indeed unless the material is finely chopped to expose the inner layers.

Proteins and **fats** are mainly found in nuts, oilseeds and animal products. Proteins, like sugars, have a capacity to bind water, but to a lesser degree. Fats reduce drying rates and thus increase drying times. For example, very oily fish dry far more slowly than those with less oil.

2.2 Public health risks of different foods

Some foods, in particular animal products, carry a much higher public health risk than others. In simple terms foods can be divided into two broad groups:

○ those with a high level of acidity and a pH below 4.5
○ those of low acidity with a pH above 4.5.

The pH scale runs from 0 to 14. Pure water has a pH of 7.0, acidic materials have values from 0 up to 7, while alkaline materials have values from 7 to 14. The pH of a substance is measured with an instrument called a pH meter, or with special papers that change colour depending on the pH.

The acidity or pH of a food controls the type of micro-organisms that can grow. Acidic foods, which include most fruits, cannot support the growth of food poisoning organisms (pathogens) such as salmonella and coliforms. If any deterioration occurs, only acid-tolerant moulds and yeasts will be able to grow which, with rare exceptions, carry no public health risk. Foods with a low acid content, on the other hand, which include vegetables and animal products, can support the growth of food poisoning organisms. Some fruits, notably banana, papaya, watermelon and prickly pear, have a pH above 4.5 and should be treated with great caution.

It is important to understand that when a food is dried, the micro-organisms naturally present on it, or picked up by contamination during processing, may not be destroyed. The drying process simply reduces the moisture content to a level at which they cannot grow. Many remain dormant and are capable of growth should the moisture content rise, for example by moisture pick-up due to inadequate packaging or poor storage conditions. The importance of good hygiene measures, such as hand washing, when preparing foods is thus obvious. While it is vital to maintain good hygienic conditions when processing any food, the level of hygiene and quality control required is higher when processing low acid foods, and in particular, animal products that by their nature carry the greatest public health risk.

2.3 Condition and quantity of air used for drying

It is commonly believed that heat is the most important requirement for drying. This is not so; the **condition and quantity** of the air used is the main driving force for moisture removal. Meat products, for example, may be dried in cool, dry mountainous areas.

The amount of water in air is referred to in terms of humidity. The actual amount of water in the air is known as the **absolute humidity**, and is usually expressed in kilograms of water per kilogram of air. Absolute humidity is not greatly used in this book, being mainly restricted to the chapter on dryer design calculations.

The more common term **relative humidity** is the absolute humidity divided by the maximum amount of water that the air could hold when it is saturated. Relative humidity (RH) is always expressed as a percentage, with fully saturated air having an RH of 100 per cent. As air is heated it expands but, it should be noted, its absolute humidity (the weight of water it contains) remains the same, as none is lost as the air expands. The relative humidity, however, falls as the same weight of water is now distributed in a greater volume. This is shown pictorially in Figure 2.1.

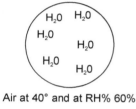

Air at 40° and at RH% 60%

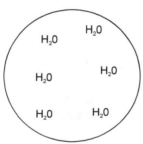

Air is heated to 60°C and expands. The quantity of water in it remains the same, but the RH% falls to 30%

Figure 2.1 The percentage relative humidity (RH%) is lowered as the air is heated and expands, while the absolute humidity (the total amount of water carried) remains constant.

The relationship between temperature, RH% and the weight of water that the air can, in theory, remove from a food to reach saturation is shown in Table 2.2.

Table 2.2 shows clearly how important the RH% of the ambient air is in any drying situation; almost 50 times

Table 2.2 The effect of heating air, initially at 29°C and 90% RH, upon theoretical water absorption

Air temperature (°C)	RH%	Amount of water theoretically removed per kg of air to reach 100% RH (g)
29	90	0.6
30	50	7.0
40	28	14.5
50	15	24.0

as much water may be removed by heating the air to 50°C. **Efficient drying occurs when large quantities of air of low RH% is available**.

In a dry, windy, cold mountain environment, for example in northern Europe or the Andes, the air may have an RH as low as 20 per cent. This very dry air, despite being cold, is capable of absorbing substantial quantities of moisture from a food without any heating. For this reason many traditional products, such as dried meats and fish, are air dried in such regions. On the other hand, the air in a humid tropical climate, with ambient temperatures of 30°C and 90 per cent RH, cannot remove any realistic quantity of water from a food unless the air is heated to lower its RH%. In such areas traditional foods are dried over fires.

A graphical representation of the characteristics of air, its temperature, absolute humidity, RH% and density, has been developed to assist drying calculations. It is known as the psychrometric chart. A simplified version is shown in Figure 2.2.

In Chapter 6 the properties of air and the psychrometric chart are described in greater detail and are used to show how an artificial dryer, fitted with a heat source and a fan, can be designed for a particular application.

The science of drying, however, is not so simple. In Section 2.4 it will be seen that it is the rate of release of moisture from the food particles being dried that ultimately controls the drying rate.

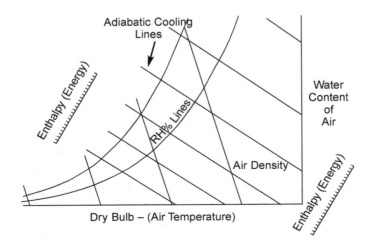

Figure 2.2 Simplified psychrometric chart.

2.4 Drying rates

When food is placed into a stream of heated air in a dryer there is a first short phase during which the surface heats (shown as A–B in Figure 2.3).

This initial settling down phase is followed by two distinct phases of drying. In simplified terms, the first phase involves the removal of water from the surface of the particle by evaporation. If the condition of the air, i.e. its temperature and RH%, are constant, the water is evaporated at a constant rate from the surface of the food. This part of the drying cycle is known as the **constant rate period** (shown by B–C in Figure 2.3).

As drying proceeds, water must be removed from the interior of the particle. This becomes more and more difficult as the moisture has to move further and further to reach the surface, where it is carried away by the air. The rate of drying begins to fall and this is called the **falling rate period**. Eventually no more moisture can be removed and the product is said to be in **equilibrium** with the drying air (C–D in Figure 2.3). During the falling rate period the rate of drying is largely controlled by the chemical composition and physical structure of the food as described in Section 2.1. Drying temperatures also become increasingly important, since heat assists the speed of migration of water to the surface.

Drying theory and the properties of air commonly appear complex to those first introduced to the subject. The phenomena described above are, however, well known and understood all over the world and can be related to everyday tasks. After washing a heavy item such as a towel, for example, it is placed outside,

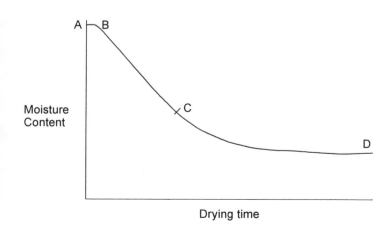

Figure 2.3 Example of a drying curve.

11

exposing as much surface as possible, and preferably on a dry, windy day (high air volume and low temperature). The final drying is carried out indoors in a warm cupboard or near a fire (low airflow and high temperature).

The drying curve data can be plotted in a different and more useful way, by plotting the moisture content against the change of moisture with time. This is known as a **drying rate curve** and an example is shown in Figure 2.4.

Drying rate curves can provide valuable information when designing dryers and planning production. Clearly, the longer the drying time, the lower the daily output in a given system. The falling rate period provides useful guidance on the correct time to remove the product as prolonged drying times, when little or no moisture is being removed, represent higher fuel costs and wasted production capacity.

Drying curves can be prepared using an inexpensive test dryer as described in Section 6.2. It is important to bear in mind that drying curves only provide **comparative or indicative information**. Actual drying times will depend upon factors such as the airflow patterns in the particular dryer being used. Some indicative rates obtained using a test dryer are shown in Table 2.3. These results clearly demonstrate the effect of cutting material into small pieces on the drying time; the pineapple rings requiring some 50 per cent more time to dry than the small dice. They also clearly demonstrate how rapidly fibrous herbs dry compared to high sugar fruits.

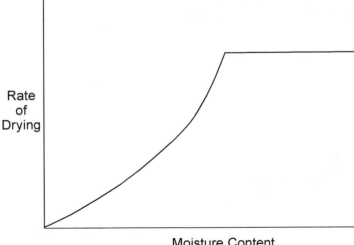

Figure 2.4 Drying rate curve.

Table 2.3 Drying times of some foods in a test dryer at 60°C

Material dried	Time to reach equilibrium moisture at 60°C (min)
Sugared pineapple dice, 5 × 3 × 1.5 cm	420
Sugared pineapple rings, 3 cm thick	600
Sugared papaya chunks, 2 × 3 × 1.5 cm	555
Banana slices, 0.5 cm thick	550
Medicinal herb – mint	160
Medicinal herb – camomile	180

Source: Hilario/ITDG-Peru 1999

2.5 Case hardening

Case hardening is a phenomenon that can cause considerable problems when drying certain types of foods, particularly fruits, certain roots, meat and fish. These types of commodities contain components that are soluble in water; sugar in the case of fruits, and minerals in the case of roots, fish and meat.

If the product is dried too quickly, soluble matter is carried to the surface where it forms a hard layer that forms a barrier to further movement of moisture from the centre of the food particle. This can drastically reduce the drying rate, indeed drying may cease. The centre of the food particles may be moist although externally the product appears dry. After removal from the dryer this moisture can allow micro-organisms to grow with resultant spoilage. It is good practice to cut samples of the food and check that the centre of the food is dry. The only way to control case hardening is to control the dryer in a way that avoids too rapid drying in the early part of the drying cycle. This can be achieved by reducing the temperature or the airflow rate. Alternatively, the air humidity can be increased by re-circulating part of the moist exhaust air leaving the drying chamber.

2.6 Particle size

As has been described, the main factor that controls drying rates during both the constant and falling rate period in the drying curve is the rate that moisture can move from the interior of a piece of food to the surface. Clearly, the shorter the distance, the faster the drying rate. For this reason, whenever possible, products should be cut into small pieces prior to drying. Reducing the size also dramatically increases the surface area. The greater the surface area, the greater the contact area

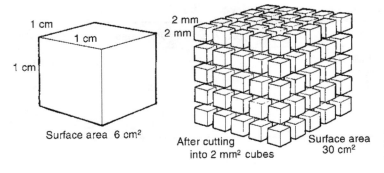

1 cm

1 cm

1 cm

Surface area 6 cm²

2 mm
2 mm

After cutting
into 2 mm² cubes

Surface area
30 cm²

Figure 2.5 The effect of size reduction on surface area.

with the air which carries moisture away from the surface. The increase in surface area resulting from cutting a 1 cm cube of food into 2 mm² cubes is shown in Figure 2.5.

Having examined the basic principles of the preservation and drying of foods, the next chapter considers the starting point for any new enterprise, the market and marketing.

3 Markets and marketing

IF A BUSINESS is to succeed, it is essential to identify or create a market for its products. Unfortunately, in too many cases, inexperienced entrepreneurs tend to rush into production assuming that a market will exist for their product. This chapter first describes how to carry out simple market research and then briefly examines issues, including the package and label, related to marketing the food.

3.1 Market research or start with the market

It should be noted that there is a clear distinction between **market research** and **marketing**.

○ Market research identifies the potential market for a product: who, when, where, what quantities and at what price.
○ Marketing promotes the product by, for example, advertising and presentation.

The aim of market research and marketing is to put in place a management process that matches the enterprise and its products to market opportunities.

Four distinct types of market exist:

○ local, within a town or limited area
○ national, throughout large parts of a country
○ regional, to neighbouring countries
○ international, to distant countries.

This chapter concentrates on local and national markets, as few small producers would be able to meet the demands of export markets. Specific issues related to regional and international markets are briefly described in Section 3.6.

3.2 Market segments

It is very important to develop the product, its presentation and method of sale to meet the needs of the

15

potential buyer. If traditional, well-known foods are to be produced, much historical information regarding likely consumers will be available. However, in the case of new or locally unfamiliar foods, little or no information will be available. The market survey must reflect the consumer's familiarity with the product. The first question to be asked is: **who are the expected customers?** These could include:

○ children
○ men or women
○ people interested in healthy foods or with special dietary needs
○ restaurants and hotels
○ institutions such as prisons, hospitals or schools
○ enterprises, such as bakeries, using the product as an ingredient
○ regional or international buyers.

Clearly the requirements of each of these groups, which are called **market segments**, will be different. Children using pocket money to buy a sweet snack will, for example, be greatly influenced by aspects such as small packs at a low price, the use of bright packaging and a convenient location to purchase, perhaps close to a school. The catering trade, institutions and enterprises, on the other hand, would not be interested in attractive packaging but would be more concerned with price, quality and reliability of supply. Importers in other countries will in general be looking for large quantities, reliability of supply and consistent quality. When exporting, it is very important that the buyer and seller develop mutual confidence.

In many cases, it is possible to break down a broad market segment into smaller, more clearly defined segments. Families, for example, can be divided into groups with different expectations and demands such as:

○ the rich
○ middle class
○ poor
○ urban and peri-urban
○ rural.

The buying habits and reasons for selecting a particular product will vary from group to group. Middle class families with two working parents would, for example, be expected to be attracted to convenience foods. Such products, however, would be unattractive to poor people whose food choices are governed largely by price. They may also be unattractive to rich families, who may have the luxury of servants to prepare fresh food. The product, its packaging, cost and method of advertising and distribution will need to be different for each of these groups.

Having considered the expected market segment or segments, it is necessary to carry out a market survey. It should be borne in mind that a given product, such as dried fruit, might have several potential market segments. For example, it may be possible to sell a dried fruit in small low cost packs as a snack food for children and at the same time market it in bulk to caterers or institutions.

3.3 Market surveys

While there are agencies that specialize in market surveys, it is usually better if the owner of the enterprise designs and carries out his or her own market survey, if necessary with some assistance from advisers. The process of carrying out the survey allows the owner of a small enterprise to gain a much deeper understanding of the customers than if the task is left to a third party.

Market research relies on two broad types of data, **primary and secondary**. Primary data is collected directly through activities, which include:

○ personal interviews in the street or shops
○ observation
○ discussions with groups of potential customers
○ questionnaires and postal surveys
○ telephone surveys
○ visiting potential buyers.

Secondary data is taken from existing reference material and includes:

○ yellow pages of telephone directories
○ trade associations and journals
○ government and official records
○ competitors' literature.

The first step in conducting a market survey is to decide what information is required. Where possible this is best done as a team, made up of partners, family members or hired help, using a structured approach as shown in Figure 3.1.

Key areas of information required	*Where and how to obtain the information*
1..	....
2..	....
3..	....
4..	....

Figure 3.1 Market survey information sheet.

Having decided what information is required and how it will be obtained, it is usual to prepare a questionnaire, similar to the example shown in Figure 3.2, to interview potential customers. Interviewing one customer will not provide reliable information. The usefulness will be greatly increased if ten people are surveyed and even more so if the sample is increased to twenty or more. As the sample size is increased, it will be found that results begin to repeat themselves. This indicates that a representative sample of customers is being surveyed.

In some countries it is acceptable to ask people if they consider themselves rich, middle class or poor. In other cultures such questions may be seen as offensive. In

Question	*Response*
1. *Do you buy snacks in the street?*	*Yes..... No.....*
2. *How often?*	*Daily..... Weekly..... Monthly.....*
3. *What kind of snacks do you like?*	....
4. *Do you think that fruit snacks are healthier than fried snacks?*	*Yes..... No..... No difference.....*
5. *Which fruits do you prefer?*	*Mango..... Papaya..... Pineapple.....*
6. *How much are you prepared to pay?*	*25c..... 35c..... 45c.....*

Figure 3.2 Typical consumer questionnaire.

general, it is important to avoid direct personal questions related to income, etc. People carrying out surveys should be trained in how to gather this type of information indirectly, in the form of an 'interested' conversation without the person being interviewed being aware. Open ended questions such as 'what type of dried fruit do you prefer?' rather than closed questions such as 'which do you prefer, dried mango or dried pineapple?' give more useful data. Typical questions could be:

○ Where do you live? (Reply will yield urban or rural.)
○ Which school do your children go to? (May provide information on class.)
○ Do you have a television set? (If yes, would indicate middle or upper class.)
○ Which shops do you use? (Richer people tend to use higher priced shops.)

The answers to the above type of questions should allow the interviewer to broadly classify the social class of the person.

3.4 Market size and value

The market survey does not produce information about the potential total market demand for a given dried food product. In order to estimate this it is necessary to extrapolate the results from the small sample of potential customers to the total market under consideration.

For example, a survey of the demand for dry onion rings, aimed at the hotel and restaurant sector, has shown that 30 per cent of owners indicated a positive interest in using the product. The next step is to determine how many hotels and restaurants exist in the city. This can be done in several ways.

Firstly, the city can be divided into areas and a physical count of all such establishments made. Alternatively, it is often possible to use secondary data such as tax information or registration of hotels and restaurants by public health bodies to obtain a full list. This type of analysis will typically give data as shown in Table 3.1.

A second example, shown in Table 3.2, is for a dried snack food. Here the market survey has examined the buying preference and habits of people from different income groups and used official statistics, taken from

Table 3.1 Total market demand for dry onion slices

Number of hotels surveyed	100
Percentage stating they would purchase dry onion	30%
Demand per month per hotel	1 to 3 kg (average 1.5 kg)
Total demand from 30 interested establishments	45 kg per month
Number of hotels registered with public health in city	500
Total demand based upon marketing survey	225 kg per month

Table 3.2 Typical market survey of a dry snack food

	Low income	Middle income	High income
Total number from local council	50 000	5000	1000
Percentage who would buy	20%	40%	60%
Possible number of customers	10 000	2000	600
Preferred pack size (g)	100	200	500
Number of likely purchases per month	1	8	6
Likely sales volume per month (kg)	1000	3200	1800
Sales value at $2 per kg	$2000	$6400	$3600

local council records, to extrapolate the small sample surveyed to the population of a town. It is of interest to note in this example that the medium income group purchases most of the product.

As can be seen from Table 3.2, producing a range of pack sizes based on the preferences of different economic groups can have a dramatic influence on total sales values. The data highlights the fact that the producer would be better advised to concentrate attention on marketing a well-presented package to middle and higher income groups.

3.5 Market share

The above examples assume that the enterprise in question has no competition. This is very rarely the case. The total market for a product, identified above, will be divided between the competition. The market share that a particular enterprise can expect to gain will depend on three main factors:

○ the number of competitors
○ the size of the competitors
○ whether the products are similar or different
○ the pricing and presentation of any competition.

This area has been studied in detail by market research specialists and the typical percentage market share that an enterprise might be expected to obtain under different scenarios is shown in Table 3.3.

This table clearly demonstrates the importance of researching possible competition from other producers. The percentage possible market share ranges from zero per cent, in the case of a high number of large companies, to 100 per cent in the case of no competitors. It should be remembered, even in the last case, that new competitors might quickly enter the market if they see an opportunity.

3.6 Regional and international markets

Processors selling to regional or international markets will rarely be able to survey the final consumers of their products. Indeed, in many cases, they will not have met the buyer. Products for export are invariably sold in large quantities to wholesalers or major companies using food ingredients. The primary processor is thus distanced from the consumer. Selling to such markets is largely governed by factors that include:

Table 3.3 Example of likely market share with number of competitors

	Likely share of market (%)
Many large competitors with:	
Similar product range	0–2.5
Dissimilar product range	0–5
Many small competitors with:	
Similar product range	5–10
Dissimilar product range	10–15
Few large competitors with:	
Similar product range	0–2.5
Dissimilar product range	5–10
Few small competitors with:	
Similar product range	10–15
Dissimilar product range	20–30
One large competitor with:	
Similar product range	0–0.5
Dissimilar product range	10–15
One small competitor with:	
Similar product range	30–50
Dissimilar product range	40–80
No competitors	100

Source: G. R. Whitby, personal communication

- product price
- meeting strict, defined quality specifications
- meeting demands such as the quantity to be shipped, frequency of shipment and timing
- providing pre-shipment samples for the buyer to examine prior to an order being placed.

While export markets may appear very attractive to producers because of the large quantities involved, some caution is required. Many large buyers, for example, are constantly searching for lower priced, better quality products. In many cases after placing one or two orders they may suddenly move to a new supplier. Simple changes in exchange rates or tariffs can trigger moves.

Distant markets, in addition to supplying a specification, will usually define how the food should be packed. The quoted price they will demand will be either c.i.f. (cost, insurance and freight) which represents the price of shipping to the destination port or f.o.b. (free on board), which is the cost loaded at the local port. Payment to the exporter is usually by means of what is called 'an irrevocable letter of credit'. This is a letter, held by a local bank, which authorizes that bank to make payment as soon as the goods reach their destination.

In most cases, export greatly depends upon mutual confidence between seller and buyer. Each has to have total trust in the other. The buyer needs to be sure that goods of the required quality will be delivered on time; failure to deliver a vital ingredient might, for example, lead to the temporary closure of a processing plant. The seller needs to be assured that the buyer will pay on time and will not suddenly source from a slightly cheaper supplier.

It is recommended that readers considering export should consult a local Government Export Agency or Chamber of Commerce. They should also be aware of subsidized opportunities that may exist for visiting trade fairs to show their products to potential buyers. Relevant trade journals should be standard reading.

In recent years there has been a growth in what are called Alternative (or Ethical) Trading Organizations (ATOs). The aim of such organizations is to benefit poor producers in developing countries. While in the past they tended to concentrate on goods such as handicrafts, some now import foods such as coffee, dry fruit, dry coconut and herbs. Such organizations do not work directly with individual entrepreneurs but with producer groups and cooperatives. Typical linkages between ATOs and producer groups are described in the case studies from Honduras and Bangladesh. The International Federation of Alternative Traders (IFAT) is based in England and the address is included in the Useful Contacts list towards the end of this book.

3.7 Marketing

Many small producers regard marketing as selling rather than a well planned management process to promote their products. Marketing requires the development of a strategy that will bring the product to the attention of the buyer, and convince retailers that it is to their advantage to stock the product and provide good shelf space for display. The aim of marketing is to convince customers that there are advantages in buying your product rather than that of a competitor. A good marketing plan can increase sales and profits and is thus a key to success. Large food companies spend considerable sums of money on advertising through papers, television and posters. Such methods are, for cost reasons, beyond the means of small producers, but low cost alternatives exist and should be used.

The first step in developing a marketing promotion plan is to clearly define who is to be targeted and how. Marketing costs should be identified and costed in the business plan. Examples of methods commonly used include:

○ advertising where potential customers, previously identified in the market survey, are most likely to see them
○ giving away free samples and attending trade fairs
○ providing retailers with an incentive for selling the products, for example a discount if highly visual shelf space is provided, a sale or return system, or rapid delivery
○ providing workers with T-shirts or hats bearing the company logo to advertise 'on the street' as they come to work
○ paying careful attention to the label design so that it catches the eye of customers, stands out from competitors' products and presents a clear message

○ placing advertisements in a newspaper or in local trade journals.

In summary, those considering establishing or developing a small enterprise based on food drying (or any productive activity) should pay serious attention to detailed planning of all aspects related to the market. Such planning should be an ongoing process as changes are continually taking place. New competitors may appear, customer tastes may change and the opening of new business opportunities may occur.

3.8 Packaging and labelling

The package and label are, to a considerable extent, an integral part of the marketing process and are therefore considered in this chapter.

Packaging

Dry food products will either be packed in bulk or as small retail packs. In both cases it is vital that the packaging protects the product from the hazards it will face until used by the customer. Common risks include:

○ Absorption or loss of moisture. In humid climates dry foods tend to absorb moisture from the surrounding atmosphere, while in dry areas they will lose moisture and dry out. The selection of the best packaging material is thus related to the local climatic conditions. It should be noted that in some cases the product may be packed in one climate, for example a dry area, and then transported to a very humid area for sale. In such cases either special materials will need to be used or, alternatively, the product will need to be marked with a shorter 'best before' date.
○ Physical damage such as crushing, breaking, loss of colour due to the action of sunlight and absorption of oxygen with resultant development of rancidity.
○ Attack by insects or vermin. Insects can easily eat their way through paper and flexible packaging.

Types of packaging material

While plastic bags and films are the most commonly used materials for packing dried foods, cardboard tubs, glass jars and aluminium foil laminates also find wide use. Flexible films differ in their ability to act as a barrier to air, moisture, light, insects, puncture and oxygen. A summary of the properties of common types of flexible packaging is shown in Table 3.4.

Plastic films can be purchased by weight in roll form or as made-up bags. If the volume to be ordered is sufficient, both can be printed with a label at reasonable cost. Most small producers prefer to buy made up bags despite the fact that they are more expensive.

After filling, the plastic bags should be heat sealed, preferably using an impulse heat sealer, so that the sealing time can be adjusted to suit the material being used. The use of cardboard boxes as an outer package is strongly recommended to avoid crushing and to protect the food from light.

Heat-sealed heavy gauge, high-density polythene is the most commonly used material for bulk packaging. As the surface area per unit weight of product is low, it provides good protection against moisture. Frequently outer protection, a sack or cardboard box, is used to provide protection against light and physical damage.

Labels

The label is a very important item and should be considered as an integral part of the marketing plan. The

Table 3.4 Properties of some common packaging films

Type of film	Properties
Polythene (low density)	Cheap, transparent, heat sealable. Poor resistance to oils, moisture and air. Easily punctured and no protection against crushing. Degraded by sunlight.
Polythene (medium density)	Reasonable barrier to moisture, heat sealable and strong. Poor barrier against odours, air and oils. Less transparent. Degraded by sunlight.
Polypropylene	Transparent, glossy, strong, heat sealable. Good barrier to moisture, air and odours. Puncture resistant.
Coated cellulose	Resistant to air, moisture, oils and odours. Heat sealable.
Aluminium foil laminates	Very good barrier properties. Heat sealable. Expensive.

labelled package is the primary contact between the producer and the buyer. It has to catch the eye, fully inform the customer what they might expect from the purchase, meet local food labelling laws and project the image that has been decided in the marketing plan.

Labels will either be printed directly onto the packaging used or, more commonly in the case of small enterprises, be applied by hand. Self-adhesive labels, applied before filling to avoid creasing, are increasingly available and are recommended. Too commonly, labels are slipped into the package before heat sealing. This is bad practice as many printing inks are toxic. It is better to seal the label into a pocket above the food as shown in Figure 3.3.

Figure 3.3 The correct way to use internal labels.

Labelling requirements vary from country to country and local advice should be sought. As a minimum the label should state:

○ the name of the product and brand name
○ the net weight
○ the ingredients, in order of amount, with the largest first
○ the name and address of the manufacturer
○ how to prepare the product if appropriate
○ the 'sell by date' and/or 'use by date'.

Increasingly, other information such as storage recommendation, and nutritional information is required.

The owner(s) of the enterprise should now have a strong indication of the types of customer that are likely to purchase the product, which will inform how the product should be packaged, presented and marketed. Market surveys will have identified the market chain and the potential monthly demand for the product. This information enables the entrepreneur to select the type and size of technology required to begin production. The following two chapters examine drying technology options and the pre-drying stages required for common types of foods together with areas that need to be considered before establishing production.

4 Small-scale drying technologies

THE CHOICE OF a particular drying system requires the careful consideration of a number of factors, which then need to be balanced according to the particular constraints that the producer faces. The principal factors are:

○ The amount of material that has to be dried, which will define the size or number of dryers required. The results of market surveys, as described in Chapter 3, will have provided indicative production rates.
○ The capital cost of the dryer and whether it will need to be imported or can be totally or largely built locally. In this respect, it should be borne in mind that in remote, rural areas items not available in the village might represent an import in the minds of local people.
○ Running costs and availability of labour, fuel and electricity.
○ The amount of value added to the food by drying. Clearly, the added value per batch must be sufficient to leave an income for the producer after all the costs involved have been covered.
○ The local climate at the time of year in which the product will be dried. Should the harvest coincide with the rainy season, for example, sun or solar drying would not be an option.
○ Social factors, including working hours and ownership issues, the latter being particularly important in co-operative or group projects involving women who may have conflicting responsibilities.
○ The type of food being dried and the quality require-ments or specifications of potential buyers. If buyers have very strict, high quality specifications, more sophisticated equipment with good control is likely to be required.

○ In addition, aspects such as the availability of good quality water for washing foods prior to drying and the availability of preparation and packaging equipment and packaging materials must be carefully considered. Simple methods of water treatment are described in Chapter 5.

This chapter describes examples of the most common types of dryers used for small-scale food dehydration. The chapter is divided into three main sections covering sun drying, solar dryers and mechanical dryers, in which fuel is burnt to heat the air and fans may be used to increase airflow rates.

4.1 Sun drying

Preserving foods by allowing them to dry in the sun or by the action of the wind has been practised for thousands of years. Today, in all parts of the world, foods such as spices, fish, grains, herbs and vegetables may be seen laid out in yards, on roofs, rocks or hanging under the eaves of houses. Such traditional drying is not restricted to tropical countries and examples of sun and wind dried products, such as fish and meat, are common in cold northern countries. Sun drying has several clear advantages:

○ Virtually no costs, except for labour, are involved, making it ideal for foods such as grains, which are simply dried for preservation with little or no value addition, or foods that are dried for home use and food security.
○ In many cases the products are dried outside, or very close to the home. The producer can therefore protect the food from theft and react quickly if a rain shower occurs.
○ As permanent structures are not required, the area used for drying returns to use for other purposes at the end of the season.

There are, however, a number of disadvantages:

○ The food is liable to contamination by dust and dirt and, if laid out close to a road, exhaust fumes.
○ Theft by birds and animals and attack by insect pests is common.

Case Study: Sun drying of maize in the Peruvian Amazon

The Shipibo Conibo people live in the Amazon region of Peru. The only access to their settlements is by river, a journey of over five hours. Until recently the Shipibo lived a traditional life based upon fishing, hunting, gathering forest products and the cultivation of cassava. Now, however, crop diversification is being promoted by the Peruvian authorities and this has led to the planting of maize.

The maize grew well in the area but the rainy season coincided with the harvest period. Pictures are available that show the Shipibo harvesting maize from boats. In some cases, due to the high humidity, sprouting has already started on the plant. It was reported that the loss of harvest was almost total. No advice or assistance was provided by the authorities on methods of drying and storing this 'unfamiliar' crop.

The Intermediate Technology Development Group in Peru (ITDG–Peru) was working with the Shipibo communities in other areas at this time and was asked to assist and quickly realized that any solution would have to be based on either sun- or fire-assisted drying followed by improved storage. While river levels rose rapidly and fields flooded during the rainy season, strong sunshine was still abundant. Trials were carried out in which the shelled maize cobs were dried on black painted galvanized roof sheets raised off the ground and angled to the sun. It was found that adequate drying was possible provided that the crop was quickly covered with polythene sheets if rain clouds approached.

A small amount of funding, provided by a Welsh church group, allowed the purchase of roof sheets, black paint and polythene sacks for storing the dry maize. These very simple interventions assisted the Shipibo reduce post-harvest losses.

This work was followed by training the community how to use maize, for example, in bakery products and tortillas.

Lessons learned:

- Through previous work in the area, ITDG/Peru had a good understanding of local social conditions and had the confidence of village elders.
- The technology chosen was simple and involved the minimum 'imports' into the communities.
- The intervention was part of a broader programme to improve the food security and nutrition of the Shipibo.

- Drying is totally dependent on good, sunny weather.
- Drying rates are slow and often the food will not be dry in one day, which means it will have stood partially dried overnight. This greatly increases the risk of spoilage through mould growth.
- It is often not possible to dry to a moisture content that is sufficiently low to prevent the growth of micro-organisms. Spoilage during storage is therefore common.

○ In the case of high volume commodities, such as rice, large areas of land are needed.
○ Some foods may darken or lose important nutrients, particularly vitamins, due to the action of sunlight.

Figure 4.1 A typical rice drying floor (© ITDG–Drick).

Despite all the above shortcomings, sun drying remains the only viable option for countless millions of small-holder producers. The products dried can be divided into three very broad groups:

○ Large volume foods such as rice, maize, sorghum and wheat. The value per tonne of such foods is low and little value is added by drying. Drying of such foods is an integral part of the production cycle to prevent post-harvest losses. In such cases an investment in improved drying technologies will not be viable unless very large-scale drying systems are used.
○ Large volume, medium value crops such as coffee and cocoa. For these crops an investment in improved dryers is often financially viable.
○ Low volume, high value foods such as fruits, vegetables, fish, meats, spices, nuts and herbs. Here an investment in more controllable drying technologies may be economically viable on a small scale, particu-larly if the food is to be packed and marketed to wholesalers or retailers.

Figure 4.2 Sun-drying medicinal herbs (ITDG–Paul Harris).

Considerable improvements to sun drying can be made by fairly simple interventions. Concrete drying floors, for example, are widely used for drying large quantities of material. This increases the rate of drying as the concrete heats up in the sun and acts as a heat store. A typical rice drying floor is shown in Figure 4.1.

When sun drying smaller amounts of food, the product should be placed on a simple table above the ground or suspended in a cloth as shown in Figure 4.2.

In some cases it may be possible to place the food on mesh trays inclined at an angle to the sun. A low-cost fish dryer, that is moved two or three times a day to track the sun, and has a rolled polythene sheet that can be quickly lowered if rain occurs, is shown in Figure 4.3. Raising the product above the ground on trays allows air to circulate above, below and through the food, thus con-siderably increasing the drying rate.

The following section describes examples of solar dryers, which can offer advantages, at reasonable cost,

Figure 4.3 Inclined fish dryer showing protection against rain (B. Axtell–Midway Technology).

over sun drying when drying low or medium volume foods.

4.2 Solar dryers

Solar dryers involve the use of a simple construction to more efficiently use the heat of the sun. Like sun drying, they rely solely on the availability of sunlight and are thus only appropriate in areas with a suitable climate: high levels of sunshine and low relative humidity.

While in many situations the use of solar dryers as an alternative to sun drying of moderate quantities of foods may appear logical, in reality the application of the technology has not been as widespread as one might expect. Solar drying has been an area of research interest in universities and institutions in most tropical countries for many years. It appears that much of this investigation has been carried out in isolation with little appreciation of the social and economic environment in which the equipment would be used. The transfer of technology to the end-user has been, in too many cases, ineffective (Brenndorfer et al. 1987).

Under the correct climatic conditions, however, the use of solar dryers can be appropriate given the advantages they possess over sun drying. A number of successful women's group drying projects have been established using small solar dryers. It is vital in such group activities to fully involve the end-users and understand the social conditions and constraints under which they live and work. It is also important that such projects are based on commercial principles, with the women for example purchasing the dryers. Commonly, aspects such as training are provided by a development agency. It is interesting to note that in almost all cases the group projects have the following common features:

○ A large number of women are involved.
○ Each woman owns a small solar dryer situated close to home so allowing her to attend to other duties while the food is drying.
○ The project is supported by an agency, which provides training and purchases the final product. The agency then checks quality and packs and markets the final product.

The principal advantages of solar dryers are:

○ Higher drying temperatures are obtained leading to both faster drying and lower final moisture contents. This means that microbiological quality and, in particular, levels of moulds and yeasts, is improved. A lower final moisture content will reduce the possibility of deterioration due to the growth of micro-organisms during storage.
○ They provide protection against rain, so avoiding the labour costs involved with quickly collecting material at times of rain.
○ They provide protection against dust, dirt, insects and theft by birds and animals.
○ They are comparatively cheap and easy to construct compared to artificial dryers, which use fans and combustion.

However, the disadvantages of using solar dryers are:

○ Unless very large they have a very low capacity and can only dry small amounts of food. This implies that little income can be generated from one small dryer each day.
○ Drying times are long when compared to artificial dryers – days rather than hours. Long drying times permit greater growth of micro-organisms.
○ Drying will stop in times of bad weather, which may result in the loss of any food in the dryer.
○ In some cases, projects have failed because raw material is not available at low enough prices.

While a wide range of solar dryer designs can be seen in different countries, all are based on two basic designs that have been modified to suit local conditions.

In the first, airflow is solely dependent on natural convection of the air heated in the drying chamber or an external collector, so causing it to move through the dryer. In the second, forced airflow, normally from electric fans, is used. Airflow rates are much higher in forced air solar dryers, leading to reduced drying times and improved final product quality. In both cases, solar air heating may be **direct**, in which the air is heated in the drying chamber containing the food, or **indirect**, using an external collector to heat the air. Some designs may incorporate some form of secondary heating by electricity, gas or burning wood to heat the air in times of bad weather or at night.

In all cases, sunlight passes through a panel glazed with transparent film, heating the air in the drying chamber or the collector. The inside of the chamber or collector should be black to absorb heat more efficiently. If black paint is used, it should be lead free and of food grade quality.

While glass is the most transparent material available, it is not recommended for glazing due to its high cost and the considerable danger of it breaking and contaminating the food. The cheapest glazing material is polythene but this tends to turn opaque and brown after a few months' exposure to sunlight. Polythene film must be replaced several times each year. A range of special plastics have been developed which are resistant to the damaging effects of ultraviolet (UV) radiation and have a claimed life of five years before deteriorating in sunlight. Unfortunately these UV-stable plastics are not yet available in many countries and have to be imported. Types of UV-stable glazing include:

○ Polycarbonate sheet, which is rigid and strong.
○ Honeycombed polycarbonate, which consists of two layers of sheet separated by a honeycomb. These are particularly efficient as they provide a considerable degree of insulation, so reducing heat loss through the panel.
○ Polyvinyl fluoride films sold under trade names such as Melinex, Tedlar, Mylar and Visqueen.

Most of the sunlight falling on the collector passes through to be absorbed on the black interior of the dryer and any food material inside. Part of the light, however, is reflected from the cover as shown in Figure 4.4.

After absorption by the blackened collector the energy is emitted as infrared radiation, or heat. It is this that heats both the air and food inside the dryer. Part of the heat absorbed by the collector will be lost through the walls of the drying chamber, hence the importance of insulating the drying chamber walls.

For maximum efficiency the collector plate should be inclined at an angle so that the sunlight strikes it at right angles, i.e. 90° (this is known as the angle of incidence). In this situation the minimum of light will be reflected

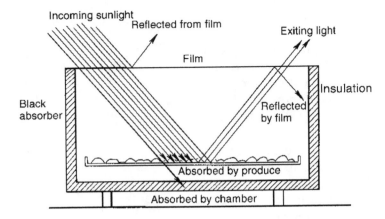

Figure 4.4 *The effect of sunlight falling on a solar collector and absorber.*

Labels in figure: Incoming sunlight; Reflected from film; Exiting light; Film; Black absorber; Insulation; Reflected by film; Absorbed by produce; Absorbed by chamber

and the maximum will enter the chamber. More and more light will be reflected from the surface as the angle of incidence increases or decreases. If the angle falls, for example to 45°, much of the light will be reflected. The angle of the collector relative to the sun is thus an important consideration in solar dryer design. In reality, a series of compromises have to be made due to the complex way that the sun moves. It rises at a low angle in the east, climbs to a peak at midday and then falls to the west where it sets. Other than on the equator, the angle of the sun's rays also varies from summer to winter. The calculation of the optimum collector angle for a particular month in a given location involves calculations beyond the scope of this publication. Interested readers are referred to the bibliography and in particular to Brenndorfer et al. (1987).

The following rules provide guidance on siting a collector. If available, expert advice should be obtained from a local institution.

○ The angle should be above 15° to allow rain water to run off.
○ The collector should be angled at 90° to the midday sun at the peak of the harvest of the food being dried.
○ In the northern hemisphere, the angle is approximately the latitude minus 23.5° in summer and plus 23.5° in winter (the reverse applies in the southern hemisphere).
○ Avoid tall trees or buildings, which may cast a shadow on the dryer.
○ Face the collector to the south in the northern hemisphere and to the north in the southern hemisphere.

Examples of ideal angles for two locations are shown in Table 4.1.

Table 4.1 Optimum collector angles

Location	Month	Angle (°)	Facing
Khartoum	April	5	South
	October	25	North
Lusaka	April	25	North
	October	5	South

Source: Brenndorfer et al. 1987

Direct convection solar dryers

The simplest type of direct dryer is often referred to as a tent dryer (Doe 1979) as shown in Figure 4.5. It consists of a frame of wooden poles covered with plastic sheet and is cheap and easy to construct. The use of black plastic on the wall **not** facing the sun improves heating efficiency. Air flows into the dryer through an opening at the base and exits from an opening at the top of the tent. The product is placed on a mesh rack above the floor. It is recommended that the poles supporting the rack are placed in tins containing kerosene, or water with a little detergent, to deter crawling insects. This is of particular importance if sweet foods are being dried.

These simple dryers provide protection against rain, dust and flies and are claimed to reduce the drying time of fish by 25 per cent (Begum 1986). They can be quickly taken down and stored until the next season or used as a temporary store. They are, however, subject to wind damage. Larger tent type walk-in dryers have been used for a range of foods including spices, fish and vegetables. They are, however, very vulnerable to damage in strong winds.

A second type of direct dryer that relies solely on natural convection is often referred to as a cabinet, Brace or Lawand dryer. It consists of a rectangular box with a hinged, glazed cover that is inclined at an angle to the sun. The chamber is normally 2–4 m long and 1–1.5 m deep, and may be constructed from wood or, as in the

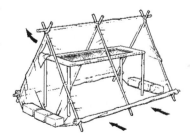

Figure 4.5 A typical tent dryer.

35

Bangladesh case study, from woven matting. A number of openings, covered with fine mesh to protect against insects, are provided in the bottom of the front wall and at the top of the back wall to allow air to flow through the chamber (Figure 4.6). In some designs air enters through the base, which is covered with mosquito netting. The typical temperature increase inside cabinet dryers is 10–15°C.

The walls of the cabinet should be insulated to minimize heat losses and the interior painted black. In Bangladesh, to reduce costs, the interior is plastered with clay mixed with soot. The food to be dried is supported on one or more mesh trays above the base of the cabinet. Attempts have been made to increase the airflow rate through the cabinet by fitting a black chimney at the rear to induce draught as shown in Figure 4.7. It has been recently shown, however, that chimneys produce little draught and that a 13 metre chimney provided less airflow than a 40 watt fan (personal communication with Hensel 1999).

Figure 4.6 Airflow in a cabinet dryer.

Indirect solar dryers

The second type of solar dryer is the indirect design and consists of a flat plate solar collector, which heats the air, connected to a cabinet containing trays of food being dried, as shown in Figure 4.8. Indirect solar dryers are more efficient and allow greater control over the drying process than direct dryers. In some cases small electric fans, powered by the mains or photoelectric panels, are used to increase airflow rates.

Figure 4.7 A Brace dryer showing chimney.

The advantages of indirect solar dryers are:

○ Products dry more quickly than in direct dryers as the air entering the cabinet is pre-heated and airflow rates tend to be higher. The quality of the final product is thus improved.
○ The size of the collector can be tailored to provide air at the required temperature.
○ Products, such as vegetables and herbs, can be protected from direct sunlight and thus retain more colour and vitamins.
○ The capacity is greater than small direct dryers and they are therefore more suitable for small-scale commercial drying of high-value crops.

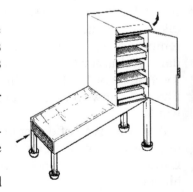

Figure 4.8 An indirect solar dryer.

Case Studies: Group use of solar dryers in Bangladesh, Honduras and Uganda

Bangladesh

In the early 1980s the Mennonite Central Committee (MCC) established an income generation project with very poor women in Southern Bangladesh. The area was rich in coconut palm but little market existed for the fresh nuts. Market studies by MCC showed that Bangladesh imported considerable quantities of desiccated coconut for use in bakeries. The studies also indicated that bakeries would be interested in a locally produced product provided that it was cheaper and of similar quality.

A Canadian food technologist was recruited and was quickly able to demonstrate that the local production of desiccated coconut was both technically and economically possible. After involving local women and undertaking training in areas such as awareness and confidence building, a pilot project – Surjonato (sunshine) Coconut – was established with 12 landless, head of family women. Coconuts were purchased in bulk by the project and the women bought nuts on a daily basis. These were opened, grated and sulphured, to retain a bright white colour, at the project centre. Each woman then took the grated coconut home to be dried in her own simple Brace solar dryer that was constructed from locally available materials (matting, timber and clay). The dry product was then brought to the project centre where it was checked for quality, packed and distributed to bakeries. The individual women were paid based on the quantity and quality of desiccated coconut produced.

By 1990 the project had grown and about 100 women were involved. It had almost replaced imported desiccated coconut in the market. However, a major problem emerged as the group could not produce in the rainy season due to the total reliance on solar drying. The local bakeries stated that Surjonato should guarantee year-round supplies or they would have to resume importation.

At this stage MCC contacted the UK-based charity ITDG and the two organizations worked together to resolve the problem. A system was developed in which water was heated by a furnace using coconut shell and husk. The hot water was pumped through a truck radiator through which a fan blew air. The heated air was passed into a cabinet containing trays of grated coconut. Economically the use of this artificial dryer was barely viable but it did allow continuous supply of desiccated coconut to the market. Despite low profits in the rainy season Surjonato was profitable on an annual basis. The artificial dryer also acted as an 'insurance' on cloudy or showery days allowing finish drying of the semi-dry product. After almost 20 years Surjonato is still in business and can be considered a sustainable success.

Honduras

In 1980 the Institute of Nutrition for Central America and Panama (INCAP) was approached by an NGO, Pueblo a Pueblo, regarding the processing of cashew nuts in Honduras. INCAP was able to provide technical assistance allowing poor farmers to process cashew nuts and Pueblo a Pueblo provided market opportunities in alternative shops in the USA.

Continued on next page

The cashew tree also produces a red fruit but this has an unpleasant bitter taste and is generally discarded. By good fortune, a scientist at the Natural Resources Institute in the UK published a paper describing a simple treatment to remove the bitter flavour from cashew fruit involving a dip in sodium bicarbonate (Ortiz 1982).

Trials were undertaken which resulted in the development of a semi-crystallized dry fruit, similar to dates. Pueblo a Pueblo then established a women's project to produce dry cashew fruit for export to alternative stores in the USA. The process involved pricking the skin of the fruit to improve penetration of sugar, followed by a dip in sodium bicarbonate. The fruit was then soaked in a strong hot sugar solution. After draining it was dried in simple, locally constructed Brace-type solar dryers. As in Bangladesh, each woman owned her dryer and the final product was taken to a centre for quality checks, packing and distribution. Payment was made based on quality and quantity provided.

By the early 1990s the project was totally managed by the 60 women involved and reports indicated that by this time some friction had developed because the women were earning more than the men who processed the cashew nuts. In the mid-1990s the project closed; the reasons are not clear but appear to relate to improved economic opportunities in the area.

Uganda

In the late 1980s Adam Brett and Angello Ndyaguma responded to a request from Ugandan farmers to find a way of processing and marketing the abundant and high quality fresh fruits and vegetables, which often rotted on the ground. They needed to find access to profitable markets for farmers to sell their produce.

It soon became clear that it was not feasible to sell fresh fruit to the lucrative export market. Fruits bruised on the bumpy roads from the villages and airfreight was so unreliable that produce would often rot at the airport. Canning and bottling was out of the question, as packaging was not locally available. What was needed was a cheap and easy way to process fruit in the villages. After investigating available options, they chose solar drying as a cost-effective process for rural based enterprises. Dried fruits have many benefits: they are easy to transport, fetch a good price in the international market and can be produced on-farm.

In the early 1990s, Adam and Angello started working with local farmers and aid agencies to develop an appropriate solar drying technology for Ugandan conditions. In 1992, Fruits of the Nile was formed as the buying, marketing and training centre for solar dried produce. Meanwhile Adam and Kate Sebag set up a sister company in the UK – Tropical Wholefoods – and began test-marketing the dried produce with health food buyers.

At first, Fruits of the Nile established only a small number of solar dryer sites. As markets in Europe developed, they were able to increase the number of people involved and today there are 60 separate solar dryer sites in Uganda, and many other farmers eager to start solar drying. Roughly 50 per cent of these are individual rural householders with small solar dryers operated by the families. Around 20 per cent have expanded to become dynamic small businesses employing local people and operating a larger number of small solar dryers. The remaining 30 per cent are women's groups and cooperatives. Each business in the latter three categories employs on average ten people in the processing of dried fruits.

When people expand their drying operations, they often incorporate photovoltaic fans into the dryer, so increasing drying rates. This also provides an opportunity to install solar lighting and power into their houses and community centres. This is an area that Fruits of the Nile would like to expand.

The enterprise has been spectacularly successful. In 1992, Fruits of the Nile bought and exported 5 tonnes of sun-dried pineapple and banana to Tropical Wholefoods. This rose to 12 tonnes in 1994, and again to 25 tonnes in 1995. In the year 2000, Fruits of the Nile exported 40 tonnes of sun-dried produce. The range of dried fruits now includes sun-dried papaya, oyster mushrooms, bananas, pineapple and chilli peppers.

The overall benefits of these locally based enterprises are considerable. A recent assessment of Fruits of the Nile estimated that at least 600 adults are benefiting from solar drying operations in Uganda. Profits from drying are spent on school fees and health as well as on improving homes and farms. As Jane Naluwairo, a Ugandan pineapple farmer, put it: 'Originally I was earning my living by running a sewing machine. I was living in a rented room with my family. When I started solar drying, I realized that it paid more than the sewing business. I mobilized some women, sold my only sewing machine and put the money into fruit drying. Now I do not regret anything. I managed to buy a piece of land, constructed a permanent house with bricks and now my family are much happier. My house has solar electricity that runs some of the dryers during the day; and during the night it provides light, and operates a television and a radio. My children's school fees are paid and their future is assured.'

One of the reasons behind the success of Tropical Wholefoods and Fruits of the Nile is that from the outset they were set up as businesses to produce quality produce for identified markets at a guaranteed, fair trade price for the producer. Care is taken to develop and provide long-term support for each link in the market chain. Thus, Fruits of the Nile does not just purchase dried fruit from farmers, it is involved in a range of activities that facilitate the production. For instance the company imports and distributes solar drying equipment and makes it accessible to small farmers; it provides effective quality controls and packaging for the European market; along with Tropical Wholefoods it provides training in all aspects of production, from building dryers to harvesting, process, packaging and transport. Another key factor in the success of the enterprise is the mutual support and dependency of all involved. If farmers experience a problem, Fruits of the Nile and Tropical Wholefoods are on hand to provide help. For their part, farmers are expected to produce dried fruits of the right quality and at the time agreed.

Local production can provide many benefits to a community. Money earned locally is spent locally and encourages the development of other enterprises. Local job opportunities help to stem emigration from the villages to the cities. In providing quality, additive-free, fairly traded produce, Tropical Wholefoods and Fruits of the Nile are helping to satisfy the demands of growing numbers of ethically minded consumers who are increasingly unhappy with environmentally unsound production methods and unfair trading practices. There is no doubt that the system developed by these companies has applications elsewhere in the developing world.

Continued on next page

Continued from previous page

Conclusions

The three short case studies show that the use of small solar dryers by groups of mainly women producers can be viable and sustainable over long periods. Several common points emerge:

○ In each case the projects were provided with technical and business support.
○ Clear market opportunities were identified by the supporting agencies prior to starting production.
○ Long-term training and support were provided.
○ Each beneficiary owned her solar dryer, which was situated at the home, so allowing other duties to be undertaken.
○ Quality assurance, packing, marketing and distribution were carried out by a central unit and not the women producers.

Indirect dryers are, however, more expensive to construct and require good carpentry and mechanical skills.

The following points should be taken into account when constructing such dryers:

○ The ground directly in front of the air entry to the collector should be covered with a plastic sheet or mat to reduce the risk of dust entering the collector.
○ The entry of the collector should be covered with mosquito mesh to stop insects entering the chamber. Similarly, all supporting legs should stand in tins of kerosene.
○ The base and sides of the collector should be well insulated.
○ Corrugated roof sheet painted black is recommended for the absorber plate as it has a larger area for heat transfer than a flat plate. It should not be painted on the underside. Alternative collectors can be made of concrete painted black, black pebbles or burnt rice husk.
○ The trays should be configured as shown in Figure 4.9 in order to promote a zigzag airflow path. This increases the contact time between the air and food and thus the rate of moisture removal. It also reduces the back-pressure that results if the air has to pass through the trays of product.

○ The outside of the drying chamber should be painted black to absorb heat.

Several institutions, including the Asian Institute of Technology in Bangkok, have investigated large indirect solar dryers. A typical example consists of a wooden frame covered with clear plastic sheet. The collector is 4.5 × 7 m, and burnt rice husk is used to provide a black absorber surface. A large chimney, covered in black plastic, was used to induce a draught. The dryer was shown to be capable of drying one-tonne batches of wet paddy rice from 20 per cent to 13 per cent moisture in two days. It was also used to dry a range of commodities, including bananas, fish and chillies (Excel 1980). While such dryers can be constructed at reasonable cost, they are highly susceptible to wind damage.

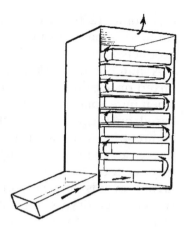

Figure 4.9 Recommended tray configuration to promote a zigzag airflow.

Tunnel solar dryers consist of a long solar collector, a drying tunnel and, in some cases, incorporate forced airflow using a small fan. The crop is placed in a thin layer in the tunnel and air, which is heated in the collector, is driven through the tunnel. Such dryers overcome the problems of wind damage associated with very large indirect dryers. Unlike the small solar dryers described earlier they are capable, under suitable climatic conditions, of producing commercial quantities of dry food. Such dryers have been promoted by a number of organizations including the Hohenheim University in Germany and the Asian Institute of Technology in Thailand.

In the Hohenheim design, the tunnel is 2 m wide, with a 16 m² solar collector and a 20 m² drying area. A photovoltaic panel, capable of producing up to 40 watts, drives a fan delivering up to 1200 cubic metres of air per hour through the tunnel, as shown in Figure 4.10. In situations where mains electricity is available, the dryer can be connected to a mains supply at night. In addition, options such as electric or gas heaters can be provided to supply heat at night or in times of bad weather. A further option, important when drying light sensitive foods such as medicinal plants, involves the use of black plastic sheeting over the product. Such dryers are reported to be capable of drying 200–600 kg of sliced fruits and vegetables per cycle, which is normally two days.

It would appear that there is considerable potential for forced-air tunnel solar dryer technology. It is claimed

Figure 4.10 A Hohenheim dryer in Sri Lanka.

that some 1200 such dryers are now operating in 55 countries, many producing commercial quantities of products, such as exotic fruits, of export quality.

A number of workers have investigated the use of mixed or hybrid solar dryers in which fuel can be burnt when pure solar drying is not possible. Many designs are based on the ideas of McDowell (1973), as shown in Figure 4.11. A firebox is used to burn wood and the hot flue gases then pass through a pipe situated under the trays of food being dried, emerging via a chimney. The layout avoids any contamination of the product by smoke. Such dryers can supplement simple solar dryers in times of bad weather or be used to continue drying after sunset. An example, constructed in Sri Lanka, used a bank of six heat exchanger tubes linked to two six-cylinder truck exhaust manifolds, to provide a greater heat transfer area and more even distribution of heat below the tray of product.

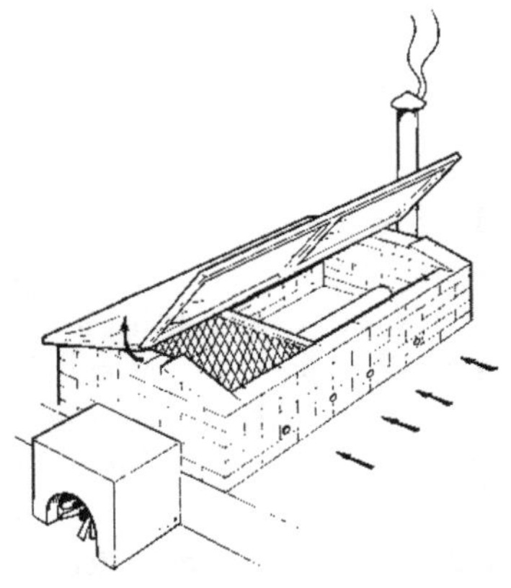

Figure 4.11 A McDowell-type biomass-assisted solar dryer.

Is the selected site appropriate for solar drying?

Without reliable data on sunshine levels, cloud cover, rainfall and relative humidity it is almost impossible to decide whether solar drying is a practical possibility or not. Such data may, in some cases, be available from a local institution or government department. If not, a system developed by the Brace Institute may provide some guidance. The system involves scoring several key questions about the local climate, as shown in Figure 4.12.

4.3 Wood and biomass dryers

A commonly expressed need in rural areas of developing countries is for dryers that:

○ have a higher capacity than small solar dryers
○ are independent of the weather.
○ do not require imported fuels or electricity
○ can dry foods more quickly.

Neither the solar dryers described in Section 4.2, nor the mechanical systems described in the following section, are able to meet these requirements. What is required are affordable, controllable dryers that preferably use

Local Climate Questionnaire

1. Do you know what the average daily radiation on a horizontal surface is, in the area where the dryer is to be set up, and for the drying period? (in kWh/day/m²)

If it is less than 2, score 0 ....
If it is 2 to 4, score 3 ....
If it is 4 to 6, score 6 ....
If it is more than 6, score 9 ....

If you do not know then answer questions 2, 3 and 4 and we will try to help you estimate it, otherwise go on to question 5 taking this value as TOTAL(1).

The following steps estimate the total solar radiation or heat available at the locality in kW/h in order to answer question 1.

2. What is the average number of hours of sunshine per day?

If less than 4, score 0 ....
If 4 to 7, score 1 ....
If 7 to 10, score 2 ....
If more than 10, score 3 ....

3. What is the length of the day in hours from sunrise to sunset?

If less than 8, score 0 ....
If from 8 to 10, score 1 ....
If from 10 to 12, score 2 ....
If more than 12, score 3 ....

4. What kinds of clouds are there?

Thick (e.g. entirely overcast), score 0 ....
Thin and misty, score 2 ....
No clouds, score 3 ....

Now add together the scores from the above three questions to give TOTAL(1).

If TOTAL(1) is:
less than 3, then availability is   2 kW/h
is 3 to 6, then availability is   4 kW/h
is 6 to 8, then availability is   6 kW/h
is more than 8, then availability is   8 kW/h

5. What is the average number of consecutive days on which the sun shines for less than four hours?

For none, score 0 ....
For one, score 1 ....
For two, score 2 ....
For three, score 3 ....
For four, score 4 ....

Continued on next page

6. At what latitude is the dryer sited?

Greater than 45°, score 0 ....
Between 30° and 45°, score 1 ....
Between 15° and 30°, score 2 ....
Less than 15°, score 3 ....

7. At what altitude (in metres) is the dryer sited?

Less than 500, score 0 ....
Between 500 and 1000, score 1 ....
Between 1000 and 2000, score 2 ....
Between 2000 and 3000, score 3 ....
Above 3000, score 4 ....

8. How can one describe the climate? Is it:

Hot and humid with very little variation in
temperature, score 0 ....
Mild, temperate and maritime, score 1 ....
Continental with large variations, score 2 ....
Desert or semi-arid, score 3 ....

9. What is the average air relative humidity during the drying
   period?

Above 85% (very humid), score 0 ....
65% to 85% (quite humid), score 1 ....
45% to 65% (quite dry), score 2 ....
Less than 45% (very dry), score 3 ....

(NB: Air with a humidity above 85% is uncomfortable and objects
stick to the skin. In a climate with less than 45% humidity clothes
etc. dry out very quickly.)

10. Are there particles in the air from dust, sand, car exhausts or
    smoke?

High concentration, score 0 ....
Low concentration, score 1 ....
None, score 2 ....

Now add up the scores of questions 6, 7, 8, 9 and 10 to give
TOTAL(2).

Finally calculate:
TOTAL(1) + TOTAL(2) − 5 = TOTAL(3)

If TOTAL 3 is:
Less than 6, the physical conditions are not right for solar drying,
other systems will have to be used to provide energy.

Between 6 and 20, another source of energy will be required as a
stand-by.

Greater than 20, conditions are all in favour of using solar energy.

*Figure 4.12 Suitability of a site for solar drying (Brace Research
Institute).*

low grade biomass such as rice husk, coconut husk or sugar cane bagasse rather than wood as fuel. While thousands of wood-fired dryers are in use around the world and many have worked for decades, little detailed information is available on designs and performance. It is suggested that this is an area that requires study by technicians and that modern technology could develop and promote efficient, controllable systems.

When using biomass as fuel it is important that the smoke produced does not contact and contaminate the food. For this reason such dryers should incorporate some form of heat exchanger which heats clean, fresh air.

A number of traditional indirect biomass dryers, known variously as Samoan or Sri Lankan, have been used for many years to dry products such as copra, cocoa, spices and fish. Most are based on a furnace-heat exchanger constructed from oil drums, which are readily available in most countries. This is situated underneath a drying platform. Hot air rises, by convection, and to a lesser extent radiation from the heat exchanger, through trays of the food being dried. In Sri Lanka, such dryers are commonly used to dry copra, using the coconut shell as fuel. The performance of such dryers is very dependent on the skill and attention of the person responsible for controlling the fire. A typical example is shown in Figure 4.13.

In Uganda a wood-fired dryer has recently been constructed and is successfully being used to dry silk cocoons. It is claimed that this dryer can attain a temperature of 100°C in the drying chamber (which is necessary to kill or stifle the cocoons) and then be controlled at 60°C for drying. The use of the dryer for foods is to be examined. Air from an external firebox passes through a heat exchanger in the form of a serpentine metal pipe and emerges via a chimney as shown in Figure 4.14. The pipe work is situated below the trays of cocoons. The firebox and ducting is then placed in a purpose-built block drying room.

In Sri Lanka the Intermediate Technology Development Group (ITDG) have developed the Anagi dryer, which is an interesting adaptation of an ITDG technology developed in Latin America, but uses rice husk as fuel rather than diesel or gas (Platt 1994). The dryer consists of a cabinet containing a stack of trays arranged to

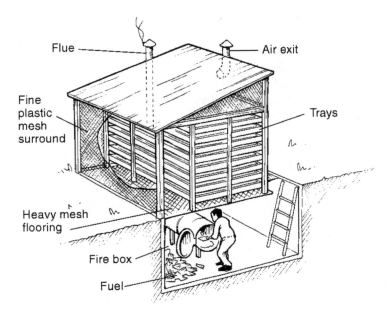

Figure 4.13 A typical Sri Lankan/Samoan dryer.

provide a zigzag airflow pattern. Heat is provided from a pot burner filled with rice husk or sawdust. Heat from the fire passes through a bank of heat exchange tubes below the drying chamber and emerges via a chimney. Heated air rises, by convection, from the heat exchanger into the drying chamber. The burner is shown in Figure 4.15. It is claimed that good temperature control can be achieved by careful control of the inlet air vents to the burner.

The cabinet holds six trays, each approximately 0.75 m² in area. A mechanical system allows the lowest tray to be removed as soon as it is dry after which all the trays can be lowered. This allows a tray of fresh material to be placed at the top of the stack of trays. The system has been very successfully applied to foods such as cashew nuts, where only a comparatively small amount of moisture has to be removed. It has not, however, proved suitable for very high moisture foods such as fruits. The main constraint would appear to be the low airflow rates achieved by convection. It is understood that consideration is now being given to increase airflow using a photovoltaic panel and a fan. Figure 4.16 shows a tray of fresh produce being loaded at the top of the stack of trays.

Figure 4.14 The heat exchanger used in the Otim dryer, Uganda.

The Lozada or Los Banos dryer (Lozada 1983) from the Philippines has been widely used to dry coconuts, maize, corn on the cob, peanuts, coffee and cocoa. The structure is very simple and can be made from locally available materials. It is claimed to have a capacity of 500 kg of corn, 300 kg of peanuts and 1000 opened coconuts. The pot burner, which is similar to that of the Anagi dryer, is fired by coconut shell.

4.4 Mechanical dryers

Mechanical dryers use a burner, normally fired by gas or diesel fuel, to supply thermostatically controlled air at a constant temperature to the drying chamber. A fan is used to increase airflow rates, so greatly reducing drying times. If diesel fuel is used, a heat exchanger is essential to avoid contamination of the food by the products of combustion. Gas burners, however, can often be used directly. Despite their higher capital and fuel costs, mechanical dryers probably provide the only fully controllable alternative to produce commercial quantities of dehydrated foods to a given specification regardless of climatic conditions. Small-scale mechanical drying systems can be divided into the following types:

Figure 4.15 The Anagi dryer burner (ITDG–Zul).

○ tray dryers, working as batch, semi-continuous or cross-flow
○ rotary dryers
○ bin dryers
○ tunnel dryers.

Batch tray dryers

These comprise a cabinet containing a stack of trays of food. Heated air, provided by a thermostatically controlled heater-blower, is ducted into the base of the chamber and rises through the product, removing moisture. An exit is provided in the top of the cabinet. A typical double chamber batch dryer is shown in Figure 4.17. The use of a double chamber offers several advantages. All of the air, for example, can be directed to one chamber when only limited quantities of food are to be dried. In addition, two different products can be dried at the same time.

As the heated air initially contacts the lowest tray, this tray will dry first. Drying then proceeds up through the stack of trays with the top tray drying last. This means

Figure 4.16 The Anagi dryer, Sri Lanka (ITDG–Zul).

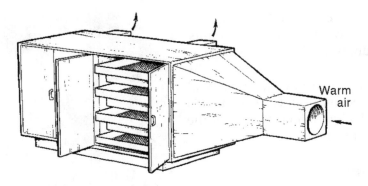

Figure 4.17 A double chamber batch dryer.

that the efficiency, in terms of energy supplied to moisture removed, falls during the drying cycle. Batch dryers become very energy inefficient towards the end of the drying cycle. Over-drying of the lower trays presents a real problem. Batch dryers, however, are simple to construct and have low operational labour costs. The cabinet is loaded with trays of product which are then unloaded after a specified drying time. They can also be left operating unattended overnight.

The energy efficiency and thus fuel consumption of batch dryers can be greatly improved by re-circulating part of the air leaving the chamber outlet back to the heater-blower. In early stages of the drying cycle, when the exhaust air is very humid, only a small percentage can be re-circulated. As the product dries, and the

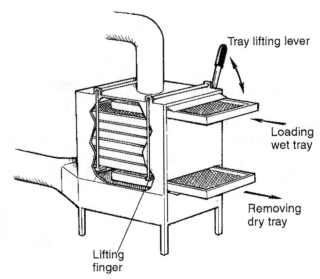

Figure 4.18 An example of an ITDG semi-continuous dryer.

humidity of the exhaust air falls, an increasing amount of re-circulation becomes possible. By the end of the drying cycle almost total re-circulation is possible.

Semi-continuous tray dryers

This type of dryer attempts to overcome the fuel efficiency and over-drying problems associated with batch dryers. Such dryers, which have been actively promoted by organizations such as ITDG, include a mechanical system, which allows the lowest tray to be removed when it is fully dry. After removal of this tray the remainder of the stack is lowered so leaving a space at the top into which a tray of fresh material can be loaded. By careful control of the weight of product placed in each tray, and the drying temperature, it is possible to 'balance' the dryer so that the air leaving the exit is of very high relative humidity and is at a temperature just above ambient, or room temperature. This means that the air has no remaining capacity for drying and the dryer is thus working at maximum efficiency. The ITDG dryer holds up to 16 trays, approximately 1×1.5 m and has a capacity of up to 400 kg fresh weight per day, in the case of herbs. A typical system is shown in Figure 4.18. Higher labour costs are involved due to the continual unloading and loading of trays. To maximize throughputs the most appropriate use of semi-continuous dryers involves 24-hour working patterns.

Case Study: Coconut processing in Vietnam

This case study is based on an article by Dan Salter first published in *Food Chain*.

In the Mekong Delta, coconuts are widely grown in small household plantations with a typical family owning less than one acre. They are an important source of income, which declined when the Socialist Bloc markets for coconut oil collapsed. The province of Ben Tre, in particular, experienced severe economic decline due to the loss of income and jobs dependent on the coconut industry. While Vietnam has successful small businesses that process coconuts for fibre and charcoal, these consume only a limited number of coconuts. Vietnam imports a number of processed coconut products including desiccated coconut.

The Swedish Red Cross promoted the planting of coconut palms to provide a natural windbreak against typhoons, which hit Vietnam's coastline annually. As a follow up to the planting programme, International Development Enterprises (IDE), a small NGO specializing in the development of market-driven, small-scale technologies, was requested to undertake a project to develop a sustainable coconut processing industry. In 1994,

Continued on next page

Continued from previous page

after extensive market research and a manufacturing feasibility studies, IDE chose des-iccated coconut as the most promising product. All desiccated coconut, widely used in the making of cookies and candy, was imported.

Over a period of two years, IDE developed a production unit to manufacture desiccated coconut and trained 80 workers and a seven-member management board in all aspects of coconut processing. Development work focused on downsizing the production methodology used in Sri Lankan factories and seeking markets in Vietnam to sustain a company. After proving the viability of producing and marketing desiccated coconut at an experimental facility, IDE established a stand-alone enterprise, the Dat Lanh Company, with some US$30 000 of machinery, much of it designed and manufac-tured in Vietnam. IDE provided a further US$10 000 of start-up capital, and the newly formed company secured bank loans of US$30 000 to renovate an existing site and upgrade services. Farmers deliver coconuts by boat to the facility where the white ker-nel is removed. Coconut shell is sold to a charcoal enterprise. The brown testa, which coats the white coconut meat, is shaved off and sold for pressing to yield a low-grade coconut oil. The kernels pass through a series of chlorinated and fresh water wash tanks before being blanched in hot water. In the drying room the kernels are ground, loaded onto perforated trays and placed in counter-current forced air dryers (similar to the one shown in Figure 4.18). Heat is supplied by a coal-fired steam boiler.

After cooling and inspection, the coconut is ground to meet individual specifications. The desiccated coconut is stored in woven sacks with heat-sealed liners. Samples of the product are taken for bacteriological testing.

The factory currently employs 80 workers, consumes 160 000 coconuts per month and pays US$190 000 per year to the 1000 to 1500 coconut-farming families who sup-ply the factory. Sales are over US$500 000 per year. The new company exported its first container of desiccated coconut to Taiwan in early 1997.

The company retains 50 per cent of profits for expansion, staff receive a percentage and 20 per cent per year is donated to support the humanitarian activities of the area.

Lessons learned:

○ The intervention addressed real local needs.
○ Sound market, economic and technical studies were undertaken.
○ The project was a partnership, bringing in external expertise as required.
○ A level of drying and production technology was chosen appropriate to the local market size and requirements.

Cross-flow dryers

Here air is passed horizontally across the individual trays of food as shown in Figure 4.19. Cross-flow dryers pro-vide further improvements in fuel efficiency and maintain a high level of control over the drying process. After pass-ing across the trays, the air is re-circulated back over the heaters. A proportion of the moist air, having passed over the trays, leaves the dryer exit and is replaced by fresh air.

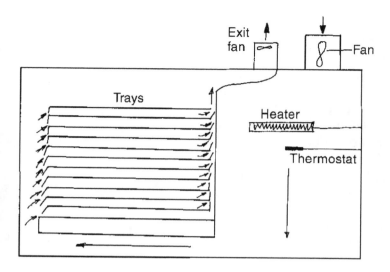

Figure 4.19 Air flow in a cross-flow dryer.

In the initial stages of drying, when the food is very moist, a large percentage of the air is exhausted. As drying proceeds this percentage is gradually reduced until, by the end of the cycle, no air leaves the exhaust and the food achieves a predetermined moisture content. Cross-flow dryers, while being highly efficient, tend to be somewhat complex and are normally controlled by humidity sensors, which operate motor driven vanes on the dryer air entry and exit.

Rotary dryers

Rotary dryers are widely used for commodities such as coffee and cocoa. They consist of a large slowly rotating perforated drum. Temperature controlled air is passed through a perforated pipe which forms the axis of the drum, and then through the tumbling mass of material being dried.

Bin dryers

These are large cylindrical or rectangular containers, with a mesh base to support the product. Low volumes (typically 0.5 m/s/m² of bin area) of warm air is passed through the bed of food in the bin as shown in Figure 4.20. The running costs of bin dryers are low. Bin dryers are used for 'finishing' the drying process by removing the small amounts of moisture from the food in its final

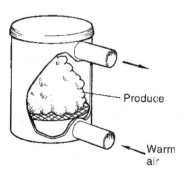

Figure 4.20 A bin dryer.

51

stage of drying, when the rate of moisture removal is very slow. This releases the principal dryer for use with another batch.

It will be found that the moisture content of individual particles of food after drying varies; some pieces will be drier than others. The final moisture content of any particular particle of food will depend on several factors. For example, larger pieces will dry more slowly. Uneven airflow distribution through trays, for example in corners, will result in pieces of food that are less dry. It is thus common practice to place products leaving the main dryer into a bin dryer in order for all the particles to reach an equilibrium moisture content.

Bin dryers also provide safe intermediate storage conditions prior to packaging.

Tunnel dryers

Tunnel dryers represent the largest technology appropriate to medium-scale enterprises. A tunnel dryer behaves in a very similar way to the semi-continuous tray dryer described earlier.

A series of carriages, each containing a number of trays of product, are passed through a tunnel, most commonly against the flow of heated air. Loaded carriages enter the tunnel at the 'wet', low-temperature end and are removed from the 'dry', high-temperature end of the tunnel. A typical tunnel dryer is 20 m long and contains 12 to 15 carriages and has a capacity of up to 5000 kg per day. Tunnels with one source of heated air often present problems due to uneven drying as the trays at the top of the carriages tend to be drier due to the fact that warm air always tends to rise. Many tunnels use systems of ducting and circulation fans to overcome this fault.

In conclusion, the range of technologies available for food drying is wide. Unfortunately, in too many cases the equipment is a prototype or is promoted over a limited area by an institution or agency. Frequently, little data is readily available on costs, capacities, suitability for different products, fuel usage, etc. For this reason readers are urged to consult both local specialists and the institutions given in the list of Useful Contacts at the end of this publication for more details and further information.

5 Processing common types of food

As described in Section 2.1, both the chemical composition and physical structure of different foods have a major influence on the rate at which they dry and the treatments required to produce high quality products. Most foods require some form of treatment before being placed in the dryer. This chapter describes processing methods that are commonly used to prepare foods for drying and briefly describes examples of the main food groups.

5.1 Principles of common pre-drying processing methods

All foods require some form of treatment prior to drying. This can range from simple cleaning and winnowing of grains or the thorough washing of herbs, to the more complex fermentation of cassava, or the salting and smoking of fish. This section examines common pre-processing methods.

Blanching

All vegetables and fruits contain specialized proteins, known as enzymes, that cause colour and flavour changes if they are not destroyed before processing. A well-known example is polyphenyloxidase, an enzyme that causes the rapid darkening of freshly cut surfaces of foods, such as apples and potatoes. A common way to destroy or inactivate enzymes is to plunge the food into actively boiling water, or to heat it in steam. This process is called blanching. Blanching also dramatically reduces the level of surface microbiological contamination that occurs during harvesting, transport and handling in the processing plant. Blanching in boiling water is simpler than steam blanching but results in greater flavour changes and losses as sugars, colours and other

components that dissolve in water are lost. Typical
blanching times required for different foods are shown
in Table 5.1. These times are only a guide and will vary
depending on factors such as the size of the food pieces,
the degree of maturity and the variety.

Table 5.1 Typical blanching times in boiling water

Food	Time (min)
Potatoes	10
Carrots	10
Cauliflower	5
Spinach	5
Broccoli	3
Celery	3
Mushrooms	2

While fruits are less commonly blanched than
vegetables, a rapid treatment is sometimes helpful when
producing products such as semi-crystallized fruits, as
described in 'Use of sugar' below.

Sulphuring and sulphiting

Sulphur dioxide or SO_2, which is either obtained by
burning sulphur or from salts such as sodium metabisulphite, also inactivates many enzymes. In addition,
sulphur dioxide acts as a preservative by inhibiting the
growth of yeasts and moulds. Sulphuring or sulphiting is
commonly used for fruits which cannot be treated by
blanching, but which tend to brown and darken during
processing or in storage. Traditionally, many small producers use lemon juice to control browning but this is
less effective than sulphur dioxide.

Sulphuring of products, such as raisins and apricots, is
carried out by burning a small piece of sulphur (about
3 g per kg of product) inside a tent or box containing
trays of food, as shown in Figure 5.1. At a large scale,
sulphuring is carried out in rooms into which SO_2 is fed
from cylinders of the gas. While sulphuring is a simple
cheap method, it mainly treats the surface with little
penetration of SO_2 into the interior of the food.

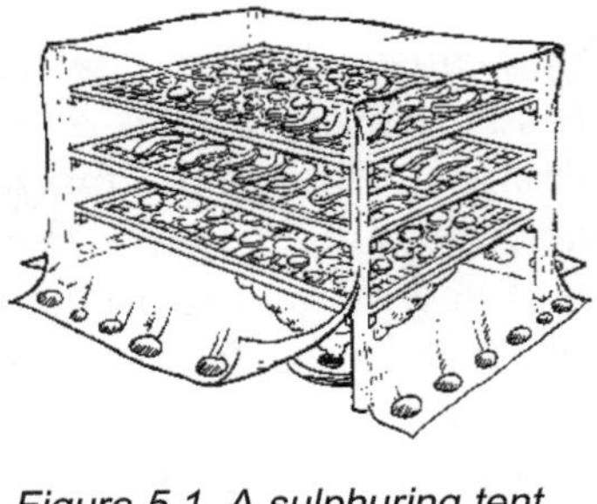

It is important to realize that SO_2 is harmful if inhaled
and can seriously damage the lungs. Operators should be
trained in its use and work in a controlled way in a well-
ventilated environment.

Figure 5.1 A sulphuring tent,
showing sulphur burning below
trays of fruit.

Sulphiting involves immersing the food in a bath or dip of sodium metabisulphite dissolved in water. If the food is to be blanched, sulphiting takes place after blanching and cooling. Sulphiting has the advantages of being more controllable, providing greater penetration into the interior of food particles and having a lower health risk than sulphuring.

Levels of SO_2 are commonly expressed in parts per million (ppm). A typical dip of 1000 ppm SO_2 is made by dissolving 16 g of sodium metabisulphite in ten litres of water. The strength of the bath and the dipping time need to be controlled and standardized to give the required level of SO_2 in the final product. This is vital if the buyer specifies maximum levels or if legislation exists regarding permitted levels. Most countries now include SO_2 under food legislation, but permitted levels vary from country to country. It is strongly recommended that advice be sought from a local standards bureau. Information on permitted levels in other countries can be obtained from the local office of the Food and Agricultural Organization (FAO) of the United Nations, or the standards institution of the importing country. Residual SO_2 in some products has now been banned in the United States of America and similar changes are expected to occur in Europe in the next few years. In such cases other control measures, such as the use of citric acid, will have to be used to suppress browning.

As a guide, SO_2 should normally be absent in foods that can be blanched, such as carrots, or be less than 200 ppm in foods that require sulphiting. Examples of permitted SO_2 levels in the UK are shown in Table 5.2.

Table 5.2 Permitted levels of sulphur dioxide (SO_2) under UK legislation

Food	SO_2 level (ppm)
Dry potato granules	400
Ginger	150
Onion	300
Mushrooms	100
Apricots, grapes, figs	2000
Bananas	1000
Coconut	50
Candied fruits	100

Source: MAFF 1995

Use of sugar

Many fruits are soaked in sugar syrup prior to drying. When a moist piece of food is placed in a strong sugar syrup, water moves from the food to the syrup. At the same time, sugar migrates into the food. This process is known as osmosis. Quite substantial amounts of water, up to 50 per cent of that originally present in the food, can be removed by osmosis in a few hours. The net effect is to reduce the moisture content of the food, which means that less moisture will have to be removed by drying. If the fruit is placed into a hot syrup the level of surface microbiological contamination will also be dramatically reduced. Obviously, sugar treatment also increases the sweetness and palatability of the dry fruit. The syrup can be re-used a number of times by making it back to strength with more sugar. It is very important, for economic reasons, to re-use syrups as many times as possible. The syrups absorb some flavour from the fruit and it is possible to use them to make other products such as fruit wines to provide additional income.

When very high levels of sugar are used, the final products are referred to as semi-crystallized fruits. If the sugared fruit is dried in a solar dryer the final product is often branded as **osmo-sol dried**, a marketing term increasingly associated with health foods. Fully crystallized whole fruits, such as small seedless oranges and figs, are very high-value products with specialized markets. They are made by allowing the fruit to soak for a few days in a series of syrups of increasing strength. The objective is to allow a slow penetration of sugar syrup with minimum breakage of the fruit cells, so allowing the fruit to retain its original shape. Typically the process starts with a syrup of 15°Brix. (°Brix is a scale commonly used to measure the strength of sugar solutions. It is measured with a small instrument called a refractometer.) After a few days the fruit is moved to a 25°Brix solution, then a 35°, a 45°, a 50° and finally to a 60°Brix solution. It is then removed, drained and dried. The economics of crystallized fruit production depends greatly on the skill of the manufacturer to re-use and blend the syrups.

Use of calcium salts

When drying some foods, for example mango and papaya, the structure tends to soften during drying. In such cases a dip for 15 to 30 minutes in a very weak solution of calcium chloride (normally 0.1 per cent) reduces this unwanted effect. The calcium binds with the pectins naturally present in the food and acts as a firming agent.

Salting

The use of salt is mainly restricted to the preservation of dry meat and fish. Salt helps to control the growth of micro-organisms which, in general, cannot grow in salt concentrations above 6 per cent. Salting also removes considerable quantities of water from the food by osmosis. Salting is carried out either by immersing the food in a brine or using solid salt, a traditional process known as kenching. For further information on salting of fish, it is recommended that books on fish processing are consulted regarding the levels of salt required and the duration of salting, as these depend upon the size and type of fish being processed. Traditionally, fish are sundried although artificial drying may be necessary in humid climates or in times of bad weather.

Smoking

While the production of smoked meats and fish lies outside the scope of this publication, it deserves mention as the preservation of the food is due to the reduction of moisture and absorption of chemicals from the smoke. These chemicals have an anti-microbiological action. Salting prior to smoking is a common process used for meat and fish and further assists preservation. Two types of smoking, hot and cold, are practised. Cold smoked foods are placed in smoke some distance from the fire and subjected to low temperatures that do not cook the flesh. Cold smoked products have a shelf life of a few days. In hot smoking the food is exposed to the heat of the fire and reaches temperatures of 60 to 80°C, which cooks the flesh. Considerable drying takes place and the final product has a long shelf life in a suitable dry climate. Smoked and salted hams for example, which are commonly produced in cool, dry mountainous areas, have a shelf life of one year or more. Readers interested in producing smoked, dried foods should consult the bibliography (particularly the UNIFEM *Fish Processing* book) and other relevant literature for details of smoker designs and processing methods.

5.2 Processing common food types

Before elaborating on the processing of particular foods groups, some basic principles that apply to all commodities are outlined.

Raw material selection is a critical first step in production, as a high quality product can only be made from first-class raw materials. Owners of enterprises should give clear guidance and training to workers regarding the quality of materials considered suitable for processing. In most cases large, evenly shaped fruits and vegetables should be used, as they will lose less flesh when peeled or otherwise prepared. Fragile soft fruits are best purchased semi-ripe and allowed to ripen in the processing room to minimize bruising and damage. After delivery it is often good practice to spray incoming materials with cold water to remove field heat.

Washing and selection is the next step in the production process. Selected raw materials should be thoroughly washed in chlorinated water before preparation. A wash solution can be made by adding two teaspoons of commercial bleach to each 4.5 litres of water. Strong bleach is dangerous and should be used according to the instructions on the bottle. All wash waters must be replaced as soon as they appear dirty. The washed raw material should then be placed in clean washed containers not, as is frequently seen, packed back into dirty field boxes.

The availability of good quality, clean water is essential for any type of food processing. In many parts of the world, and particularly in rural areas, water quality cannot be relied upon. Provided that local water is of reasonable quality, various low cost measures can be taken to treat it which include:

○ Boiling the water and allowing it to cool. This is, however, costly in terms of the energy required.
○ Filtering through a sand bed or an in-line ceramic microfilter cartridge.
○ Treating with chlorine (two teaspoons of bleach per 4.5 litres of water).
○ Constructing a high-level water tank with a sloping base to allow it to be completely drained. The water take-off pipe should be 15 cm above the base. In use, the tank is filled with water at night and bleach added as appropriate. The water is allowed to settle overnight

58

for use the next day. The whole tank is completely drained prior to refilling.

Good hygiene in the processing area is vital and discussed more fully in Section 6.1 and in the references listed in the bibliography. Some foods, particularly sweet fruits, attract insects such as ants. It is recommended that the legs of preparation tables are stood in tins of kerosene or detergent to deter these pests. As a general rule, the floor of the processing area should be kept wet to deter insects and reduce dust. At the end of the day all equipment, floors and surfaces should be thoroughly washed with chlorinated water. Refuse should be removed quickly from the site and properly disposed of – not, as is too often seen, thrown into an open bin just outside the processing room.

Drying ratio is a term frequently quoted in literature on drying, and defines the weight of raw material required to produce 1 kg of final product. Drying ratios provide a useful guide when calculating the amount of raw material that has to be processed for a given, planned daily output. They can also be used to quickly calculate final product costs. Two types of drying ratio are used:

○ The overall or crude drying ratio is the weight of incoming material required to produce 1 kg of final product.
○ The prepared drying ratio is the weight of peeled, or otherwise prepared material that is needed to produce 1 kg of final product.

In the case of sweet bell peppers, for example, the overall drying ratio is 22:1 and after preparation and removal of the core and seeds the prepared drying ratio falls to 14:1. This means that 220 kg of raw material would have to be purchased in order to produce 10 kg of final dry product and that a dryer with a capacity to dry 140 kg of prepared pepper would be required.

The following sections briefly examine the preparation of dry products from common types of foods.

Starchy foods

Potatoes
The Irish potato, which originated in Latin America, has been traditionally sun-dried for thousands of years. *Papa*

seca, which consists of dried pieces of partially cooked potato has, for example, been found in ancient Inca tombs. While traditional products are still produced both for home food security and income, they are becoming less popular and often regarded as 'food for the poor'.

The principal commercially dried potato products are thin slices, cubes and granules. Considerable quantities are used in products such as dry soup mixes in Europe and North America. These convenience foods are becoming increasingly popular in many developing countries as social structures change. Their production may provide opportunities for medium-sized producers.

Selecting the best locally available raw material can often mean the difference between profit and loss. Large producers select, with great care, the varieties purchased for processing which should be high in starch solids, low in sugar and with the required texture.

Very simply, potatoes can be divided into two types. The first are known as *floury* and tend to soften and break up on cooking, making them ideal for mashing. Floury types are ideal if granules or powders are to be produced. The second type can be described as *waxy*. These potatoes hold their shape well on cooking, making them ideal for chips and slices. Waxy varieties tend to be lower in starch solids and higher in sugars. The higher sugar content can have a negative effect on drying rates due to case hardening and is more prone to discoloration through browning. If diced or sliced dry potato products are to be manufactured, waxy varieties should be used for the product to maintain its shape. Some varieties of potatoes have a tendency to darken even after cooking. In some cases they blacken and should be avoided.

Exposure of potatoes to light after harvesting results in a greening of the skin and flesh. This is due to the production of a bitter toxic substance called solanine, which can pose a serious health hazard. Any **potatoes showing signs of greening should be rejected** at the washing and selection stage.

A final and important criterion when buying potatoes is the size and evenness in shape. Peeling losses will be very high if small, irregular potatoes are used, as illustrated in Figure 5.2.

It is strongly recommended that those contemplating producing dried potato should buy samples of different varieties from local growers in order to carry out simple

Figure 5.2 The importance of buying large round potatoes.

cooking, darkening, drying and 'yield-after-peeling' tests. This principle should be applied to all the foods described in this chapter. A typical flow diagram for processing potatoes is shown in Figure 5.3.

After thorough washing in chlorinated water and the rejection of any unsuitable raw material the potatoes are peeled. While the peel of a potato accounts for only 4–5 per cent of the total weight, actual losses through peeling can be as high as 25–30 per cent. In terms of product cost, peeling is therefore an area of great importance. The four methods available are:

○ hand peeling
○ mechanical abrasive peeling
○ chemical or lye peeling
○ flash steam peeling.

Hand peeling is the most costly in terms of labour and peeling loss. The capital cost in terms of equipment is, however, minimal.

Abrasive peelers consist of a rotating drum lined with an abrasive coating, somewhat similar to very coarse sandpaper. Water flows continuously through the peeler, washing away the skin as it is removed. Abrasive peelers work best if even, round potatoes are used. Irregularly shaped potatoes will require considerable hand finishing. Small commercial electric machines, of the type commonly used in hotel or institutional kitchens, are reasonably inexpensive. It should be noted that substantial quantities of dilute contaminated wastewater will have to be disposed of. Waste disposal must therefore be considered carefully in any planning process. A typical abrasive peeler is shown in Figure 5.4.

Manual and powered abrasive peelers have been constructed in local workshops from a small metal drum coated internally with an abrasive material.

Lye peeling involves dipping the potatoes in a 10–20 per cent solution of caustic soda (sodium hydroxide) at 60–80°C for 2–6 minutes. The actual conditions used are determined by trials. After this treatment the potatoes are washed thoroughly in cold water to remove lye from the surface, and the peel is removed with stiff brushes. Any remaining skin and eyes in the potatoes are then removed by hand. While lye peeling is more economical in terms of labour costs and peeling loss, it is a process

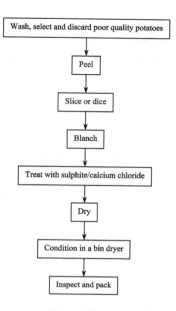

Wash, select and discard poor quality potatoes
↓
Peel
↓
Slice or dice
↓
Blanch
↓
Treat with sulphite/calcium chloride
↓
Dry
↓
Condition in a bin dryer
↓
Inspect and pack

Figure 5.3 Typical flow diagram for sliced or diced potatoes.

Figure 5.4 A small commercial abrasive peeler (courtesy of Hobart Manufacturing).

61

that requires considerable control if accidents are to be avoided. Lye, if splashed into the eye or onto the skin, **can cause blindness or serious burns**. It is strongly recommended that producers unfamiliar with the process avoid lye peeling unless expert guidance and training in safety can be provided. Ideally the tanks used for lye peeling should be constructed from stainless steel. Steel tanks will corrode and may cause staining of the product. **Under no circumstances should aluminium be used** as it will quickly be eaten away by the lye. Lye peeling is now not commonly used commercially.

Figure 5.5 A manual slicer and chipmaker. The chips produced can quickly be cut into dice with a knife.

Flash steam peeling is now the accepted method. The potatoes are placed in a large pressure cooker, which is brought up to pressure. The pressure is then suddenly reduced, causing the skin to lift away from the flesh. The skin is finally removed by high pressure water sprays and brushes.

Slicing and dicing can be carried out by hand or mechanically. Small hand operated equipment is available and often used by hotels and restaurants. Figure 5.5 shows a manual slicer and a chip press that can be used to make diced potato.

At a larger scale, powered equipment is available. Slicing machines consist of a rotating plate with a sharp adjustable knife attached to the face. The required slice thickness can be obtained by adjusting the distance of the knife from the faceplate. Slicers are of reasonable cost, and in many countries can be locally constructed. Machines that combine slicing and dicing, of the type shown in Figure 5.6, are available but expensive.

Blanching and sulphiting are necessary as some varieties of potatoes are particularly prone to enzymatic darkening, so it is important that they are rapidly blanched or sulphited after slicing. A typical sulphite dip consists of 7.5 g of sodium metabisulphite dissolved in 10 litres of water. This gives a 500 ppm (parts per million) dip, which is sufficient to control darkening and leaves residual levels of less than the legal maximum. Underblanched potatoes develop a white, chalky appearance as the product is dried. If this is detected, blanching times should be increased.

Figure 5.6 An electric slicer/dicer with changeable plates and cutting blades (courtesy of Hobart Manufacturing).

Drying at small- and medium-scale will be carried out in either a solar or mechanical tray dryer of the type described in Chapter 4. A typical crude drying ratio for potatoes is 8:1.

It is important to spread the slices or dice so that that they are not piled on top of each other, as this will slow down drying rates and may result in some pieces that are insufficiently dry, with the danger of mould growth after packing. The product should be mixed or turned at intervals to expose new surfaces for drying. The final moisture content should be 6–8 per cent.

Bin drying If possible, the dried potato should be finish dried at 40–60°C in a bin dryer, as described in Chapter 4, to a moisture content below 8 per cent to ensure that all the pieces are fully dry and at equilibrium.

Milling In some situations there may be a market for potato flour which can be reconstituted or used as a food ingredient. Potato flour can easily be produced by milling dry potato slices in a hammer or disc mill followed by sieving.

Packaging The final product should be packed and heat sealed in heavy gauge polythene bags.

Sweet potatoes and cassava
Sweet potatoes and cassava are, in terms of tonnage, the most important root crops and are grown in over 100 tropical countries. Sweet potatoes can be processed in a similar way to Irish potatoes and are usually marketed as dry slices or as flour. Many attempts have been made to use sweet potato flour to replace up to 20 per cent of wheat flour in bread making in order to reduce costs, but the results have not been encouraging.

After washing, the sweet potatoes are peeled, using one of the methods described for Irish potatoes, and held in a bath of dilute salt, citric acid or sulphite to retard darkening. They are then diced or sliced, blanched for about six minutes, dipped in a sulphite bath and dried on trays at 60–70°C.

Cassava, unlike roots such as potatoes, cannot be stored for long periods after harvesting and requires rapid processing. Cassava contains toxic cyanide compounds and special processing techniques are required to produce foods safe for human consumption. The levels of hydrogen cyanide in cassava vary greatly and range from 10–450 mg per kg of fresh root. Cassava with low levels of cyanide is often referred to as sweet while that with high levels is called bitter cassava. It has been shown, however, that there is little clear link between the degree of bitter taste and cyanide levels. Considerable caution is thus required when processing cassava for human consumption, to ensure that the correct varieties are used.

A wide range of traditional dry cassava products are produced, one of the best known being gari in West Africa. Gari is produced by allowing finely grated, washed and peeled roots to ferment for three or four days, during which time almost total breakdown of the cyanide complex occurs. Water is then squeezed from the mash using a press. Finally the cassava is placed in a large shallow pan over a fire and dried, or toasted, while continually stirring the mass to a moisture content below 12 per cent. Drying in this way on hot metal plates is widely practised in many countries.

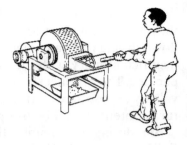

Figure 5.7 A cassava chipper.

The principal dried cassava products are chips, pellets and starch. Cassava chips are made by thinly slicing peeled roots using a rotating disc cutter fitted with cutting blades as shown in Figure 5.7.

Where the climate permits, the thinly sliced chips are spread out on drying floors or tables and sun dried, with regular turning to assist even drying. Drying to a moisture content of 13–15 per cent is normally accomplished in two to three days. It has been found that considerable drying of cassava slices can take place even at night if the product is placed on trays inclined at an angle to night breezes. In damp, rainy areas some form of artificial heat is required for complete drying.

In many countries starch is extracted from cassava for use in both food and non-food products. The washed peeled roots are finely grated and then 'kneaded' under water, either by hand or mechanically, to release the starch. The milky starch mixture is then left to settle in tanks allowing the starch to sink. The water is removed and the starch broken up into lumps and sun dried. After milling and sieving it finds use in a wide range of industries including bakeries, confectionery and adhesives.

Carrots

Carrots, in the form of dry dice and flakes, are produced in large quantities and find use in a wide range of food products such as dry soup mixes. The variety selected should have a strong red colour, be even in size and not large and 'woody'. The crude drying ratio is 15:1.

After selection and washing, the carrots are peeled, normally using lye or flash steam methods. Hand peeling is very time consuming and results in high losses. Small manual peelers, as shown in Figure 5.8, are available at reasonable cost. The peeled carrots are then diced, sliced or shredded. After hot water blanching they are sulphited and dried to a final moisture content of 5 per cent. Dry carrot tends to deteriorate after about three months, losing colour and developing an undesirable odour. Light speeds up this deterioration, therefore the dried carrots should be stored away from light.

Plantains and green bananas

These are dried in many countries either directly in the sun or by using a dryer. Both whole bananas and slices are produced. Sun-dried whole ripe bananas have a limited market as a health food in Europe and North America. This dark brown product, while not looking particularly attractive, has a pleasant fruity sweet flavour. In some cases the bananas are cut lengthways to expose more surface area, and sulphured before drying, This product is lighter in colour.

Semi-ripe bananas and plantains, cut into thin slices, are also commonly dried. The use of sulphite or citric acid is recommended to prevent darkening. Dried plantains, due to their cheapness, have been used to prepare low cost nutritious foods for schools and programmes to provide better nutrition to children. The dry slices are milled, sieved and mixed with nutritious materials such as soya flour and skim milk powder. In some cases flavours, such as cocoa powder, have been added to improve acceptability. Products of this type may offer opportunities for small-scale processors in countries where subsidized feeding programmes operate.

Vegetables

Green leafy vegetables

Green leafy vegetables, such as cabbage and spinach, are commonly dried to preserve them for use after the harvest period. After washing, any outer spoiled leaves and the woody core are removed. The overall crude drying ratio of cabbage is typically 17:1.

The leaves are then finely shredded to a width of about 5 mm. A rapid blanch in boiling water for 1–3 minutes is recommended to reduce surface microbiological contamination and help retain a fresh dark green colour in the final product. The blanching time must be

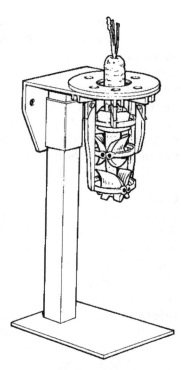

Figure 5.8 A manual carrot peeler, with a capacity of 100 kg per hour.

carefully controlled to avoid overcooking, as the leaves will become slimy and almost impossible to dry. Immediately after blanching a dip in cold sulphite solution causes rapid cooling, so avoiding over-cooking, and assists in the preservation of a fresh colour. Drying is then carried out to 5 per cent moisture at a temperature of 50–60°C. Dry leafy vegetables are hygroscopic, i.e. they rapidly absorb moisture from the air. After leaving the dryer and cooling they should therefore be packed as quickly as possible in heat-sealed plastic bags.

Onions

After potatoes, onions are the second most common dehydrated vegetable. In many countries, fresh onions are relatively cheap at harvest time but prices then rise and they become expensive prior to the next harvest. Drying during the harvest peak, for sale as prices rise, thus presents an opportunity for small enterprises or offers increased food security within the home. The growth of fast food restaurants in cities can represent an opportunity for selling conveniently prepared onion slices. Dried onion finds a use in many products such as sauces, pickles and soup mixes.

Large, strong flavoured onions are preferred for processing. These have a crude drying ratio of 10:1. After peeling and washing, onions should be evenly sliced in to rings about 3 mm thick. Onions do not require sulphiting but in some cases a dip in a 5 per cent solution of common salt can prevent browning during drying. High drying temperatures should be avoided as this encourages browning. After drying to 5–7 per cent moisture, and cooling, the hygroscopic product should be packed as soon as possible.

Mushrooms

Dried mushrooms are high value products, making them an ideal product for small-scale processors. In many countries they have a ready market both for household and commercial markets. A major problem with mushrooms, due to the way they are grown in compost, is their very high surface microbiological contamination. Very thorough washing in chlorinated water is recommended to remove all soil and compost. After washing they should be carefully inspected and any mushrooms not completely clean returned for further washing.

Traditionally, mushrooms have been dried whole. The first stage of the process is to cover the mushrooms with salt and leave them for twelve hours. This process removes a considerable amount of water by osmosis. The mushrooms are then threaded on a string for air drying. The majority of dry mushroom is now produced in the form of slices or small chips.

Prior to drying, blanching for 2–4 minutes is recommended to reduce the level of microbiological contamination. Blanching does, however, reduce yields as soluble matter is removed. Mushrooms may be sulphited to improve the colour. They should be dried to a final moisture content of 5 per cent or below.

Tomatoes

Tomatoes are traditionally dried in many countries and in recent years sales of sun-dried tomato products in Europe have grown considerably. Dry tomatoes are sold in rings, as a coarse powder, or bottled under oil, and find use in a range of dry food products. Bright red varieties with a high solids content should be used. The crude drying ratio is 22:1 or more. After washing, the tomatoes should be evenly sliced, loaded onto trays in a single layer and sprayed with a sulphite solution to avoid handling damage. They are dried to 4–5 per cent moisture. Dried tomatoes are extremely hygroscopic and must be packaged quickly.

Bell peppers

Bell peppers are a medium-value food that is commonly dried. The crude drying ratio is typically 18:1. After washing the peppers are cut in half and the seeds and heavy white internal ribs removed. They are then sliced or diced, sulphite dipped and dried at about 60°C to a moisture content of 5 per cent.

Fruits

While dried fruits such as raisins, apples and apricots dominate the market in terms of world trade there has, in recent years, been considerable growth in the market for dry 'exotic' fruits. Banana, pineapple, mango and papaya are the most important. A study in 1994, however, showed that still only about 30 000 tonnes of dry tropical fruit were traded out of a total global dry fruit market of some 500 000 tonnes (International Trade Centre 1994). The market is thus small and a high percentage of the trade is in dry banana chips for use in breakfast cereals. A British importer estimates that only about 30 tonnes of dry pineapple, mango and papaya are sold in the UK each year. The majority of dried tropical fruits traded find markets in health food shops or in products promoting health, such as breakfast cereal mixes. With increasing public concern over pesticide residues in foods it is possible that niche markets could be developed for producers able to guarantee that the fruits used are organically produced.

As mentioned earlier in this chapter, the high sugar content of fruit makes it particularly attractive to insects and measures must be put in place to deter them.

The softness of many types of fruit, such as ripe mangoes and papaya, makes them extremely liable to bruising if poorly handled. Bruised areas rapidly begin to rot and decay will spread to adjacent fruit. If possible, all fruit should be laid out in a single layer in the shade to ripen fully.

Figure 5.9 A manual peeler/corer for pineapples (ITDG–Peru).

As fruits are acidic, it is very important that only stainless steel knives are used for peeling and preparation. If ordinary knives are used they will stain the fruit. Food grade plastic containers, with lids, should be used to hold prepared fruit. It is recommended that plastic mesh is used on drying trays, again, to avoid staining. It should be noted that **pineapples and papaya contain enzymes that can damage the hands**; in fact commercially available meat tenderizers contain these enzymes. Workers should thus wear plastic gloves when preparing these fruits as prolonged exposure can result in serious damage to the skin. Pineapples require coring after peeling. A simple hand operated cutter, which removes most of the external skin and the core from pineapples, is shown in Figure 5.9.

After peeling and coring the fruit is sliced, either by hand or mechanically. The thickness of slicing can be critical. Very thin slices tend to stick to the mesh of drying trays, while slices that are too thick dry slowly and unevenly and case harden, appearing dry on the surface but remaining very moist internally, with the danger of deterioration after packing. Typical recommended slice thickness for common fruits are given in Table 5.3.

Mechanical slicing will produce slices of uniform thickness and is recommended. Again simple machines, can be constructed to rapidly produce slices with an even thickness. A cutter for slicing pineapples is shown in Figure 5.10.

Table 5.3 Recommended slice thicknesses for some common fruits

Fruit	Recommended thickness (mm)
Papaya	2–3
Mango	2–3
Pineapple	2–3
Tomato	3–5
Banana	5

Source: NRI 1998

A dip in sodium metabisulphite solution, as described in the previous section on vegetables, is recommended where its use is permitted. Sulphiting greatly assists in maintaining a bright, fresh colour and helps retard the growth of yeasts and moulds during drying.

Drying trays should be evenly loaded, avoiding laying slices on top of each other, which causes them to stick together and results in the creation of uneven damp areas. Fruits that have not been sweetened with sugar should be dried to a final moisture content of 12–15 per cent and packed as soon as they have cooled.

Semi-crystallized, crystallized and osmo-sol fruit

The flavour and acceptability of dry fruit can, depending upon local tastes, be improved by soaking the slices or pieces of prepared fruit in sugar syrup prior to drying. The use of a hot syrup will greatly reduce surface microbiological contamination. In addition to removing substantial amounts of water from the fruit by osmosis, the sugar binds moisture, so making it unavailable for the growth of micro-organisms such as yeasts and moulds. This means that the final product can have a higher moisture content without deteriorating. Typically, sugar treated fruits have a final moisture content of 20–30 per cent.

After leaving the dryer, products should be inspected, any sub-standard material removed, and then allowed to cool. The product should be packed in heat-sealed plastic bags or other airtight containers that will protect it from moisture pick-up in a humid climate, or drying out in a dry climate. It should be remembered that sweet dry fruit is very attractive to insects and all finished products should be well stored (off the ground in a cool place).

Herbs and spices

Herbs and spices are very high value commodities commonly grown by smallholder farmers. The crude

Figure 5.10 A manual pineapple slicer (ITDG–Peru).

drying ratio is approximately 8:1. They are traditionally dried by either sun-drying on mats or, in the case of herbs, hanging bundles in the shade. Post-harvest losses, due to bad weather after harvest, can result in serious financial loss for small farmers. In addition, the better quality that can be obtained if artificial drying is used can result in improved markets and higher prices. This is particularly true of culinary herbs and medicinal plants, which have a much fresher, green colour if dried quickly.

Herbs and spices require only basic, simple treatment prior to drying. They are never blanched as they contain volatile oils, known as essential oils, which provide their characteristic odour and flavour. Spices should be winnowed, either using traditional winnowing trays or mechanically, to remove dust, chaff, leaves and stones prior to drying. Certain spices require specific pre-drying treatment, for example removal of infested nutmegs by water flotation, and specialist literature should be consulted as appropriate.

When processing herbs, all thick stems are generally removed before they are dried. Very efficient washing in chlorinated water is important to reduce the level of microbiological contamination. In some cases, for example leafy herbs, buyers may define a maximum percentage of stem materials. In such cases it may be necessary to pass the dried herb through a winnowing machine to separate leaves from stems.

After drying, the final product should be packaged as appropriate to the market; retail or bulk.

Nuts

Nuts, such as brazils, macadamia, peanuts and cashews, are high value products that, in many cases, provide an important source of income to poor people. They are traditionally dried on mats in the sun and inclement weather can result in post-harvest losses due to mould growth. Nuts, and in particular peanuts, are commonly attacked by a mould that poses a serious health risk. The mould *Aspergillus flavus* produces a toxin called aflatoxin that can cause cancers if consumed over a long period of time. The mould, which is black in colour, will grow on nuts that have not been properly dried and shows as dark stains on the nut surface. It is very important that shelled nuts are quickly and fully dried after harvest. Many nuts, such as brazils or cashew, are destined for export markets and overseas buyers will test for the presence of aflatoxin, rejecting any contaminated material. If adequate sun-drying cannot be assured, artificial dryers are essential to produce high quality products. The Anagi dryer, described in Chapter 4, is being successfully used to dry cashew nuts in Sri Lanka.

Meat and fish

Meat and fish products present a far greater public heath risk than any of the food groups described previously due to the food poisoning micro-organisms they can support. Despite this, traditional methods of drying meat and fish, often combined with salting or smoking, are safely practised all over the world. It is very important to appreciate that safe foods have been developed from high-risk raw materials by selecting processing methods that take into account the local climate, social customs and local knowledge that has evolved over centuries. The greatest danger, if contemplating drying animal products, is to copy a traditional product and method from another area. For example, dry game known as biltong is produced in very arid areas of southern Africa. Attempting to produce biltong in a slightly more humid climate would expose consumers to grave risks of food poisoning.

It is recommended that those contemplating producing dry fish or meat should base their production on local, traditional, well-known products. Improvements, such as more hygienic handling, should be made as required, but changes should be considered with care and caution. Packing, for example, in a plastic bag might at first sight appear an improvement over selling dry fish from baskets. Experience has shown, however, that packing in this way may allow condensation of water on the bag's inner surface when the product is placed in the sun. The moisture droplets then fall onto the fish, causing a damp spot in which bacteria can start to grow.

Having established the potential market for the selected dried food, and identified the most appropriate pre-drying and drying systems that will be required to process the planned output, it is necessary to develop a business plan. In the next chapter issues related to planning production are described, together with guidance on preparing a business plan.

6 Planning production

W HEN PLANNING PRODUCTION a series of areas need consideration. These include the design of the production room or facility, including aspects related to safety, the design and costing of quality assurance systems and the development and costing of the product range. This information is essential for the development of a business plan and will be required if finance is sought from a bank or agency.

6.1 The processing plant

This section provides basic guidelines for the design and layout of a room or building for the safe production of dry foods. In the majority of cases, small- and medium-scale producers will use an existing building that has to be adapted for production. Attention to detail and careful thought can, at little extra cost, result in a building that is appropriate for the safe processing of foods. The first step is to ensure that the building selected is in the correct location, in a clean area away from dust, refuse, dirty drains, etc. If solar dryers are being used it is important to check that large buildings or trees do not cast shade on the area. The producer also needs to ensure that the labour force will be able to get to work easily and that adequate services (potable water, power, refuse disposal and drainage) are available.

The external appearance of the building, and its surroundings is very important. It should look clean and fresh to give customers confidence in the products. A well-made nameplate is important. The provision of toilets, as discussed in the section on worker hygiene, requires consideration. Internally, the building should be divided into two basic areas; a wet area for the preparation of incoming material and a dry area for drying, packing and storage. It is often convenient to carry out selection, washing and peeling, etc. under a covered lean-to roof. This reduces contamination levels in the processing plant. Consideration should be given to the following aspects:

○ The joint between concrete floors and walls should be curved so that it can be easily cleaned.
○ Concrete floors should be sloped to a drain to assist cleaning.
○ Windows and other openings should be covered with fly-proof mesh.
○ Drains offer an entry point for rodents and should be covered with strong mesh. Rodents can also enter buildings via overhead cables and these should be fitted with a circular metal disc.
○ Where mechanical dryers are to be used, good ventilation is essential.
○ As far as possible the material being processed should 'flow' through the plant to avoid cross contamination.
○ Finished goods should be stored, off the ground on pallets, in a closed storeroom and removed on a 'first in, first out' basis.

Further information can be found in books listed in the bibliography, particularly those related to quality assurance.

6.2 Product development and market samples

In many cases small producers will find it necessary to undertake work on new product development. Commonly, this starts by buying and examining samples of existing brands and trying to match their characteristics. Careful examination of aspects such as the colour, sweetness, texture, ingredients list and type of packaging will provide useful information.

Buyers of a new product generally wish to see and taste the product and be given indicative prices. This clearly poses a problem if no facilities are in place to produce samples. It is possible, however, with very basic equipment such as a small low cost test dryer, a domestic heat sealer, scales, knives, etc., to produce samples for demonstration with a reasonable estimate of their price. A typical test dryer and an example of a test-drying curve are shown in Figures 6.1 and 6.2.

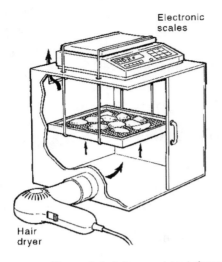

Electronic scales

Hair dryer

Figure 6.1 A low-cost test dryer.

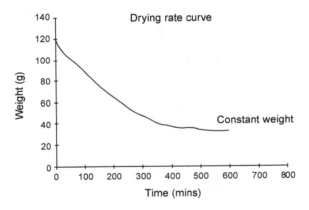

Drying rate curve

Constant weight

Figure 6.2 Test-drying curve for banana.

The test dryer consists of a plywood box with several trays supported on wood runners and a door to the front. The chamber is heated by a domestic hair dryer that has been rewired, with an external thermostat (mounted in the chamber), so that the fan runs continuously while the thermostat controls the heating elements. Test dryers of this type have been used successfully, in several countries, to develop products. Energy use can be measured by noting the percentage of time, over a five minute period that the heaters are on.

6.3 Costing the product

Product costing is an area that frequently poses a problem for entrepreneurs who are entering business for the first time. The following section briefly describes the principles involved but it is recommended that advice should be sought from local institutions, a bank or consultant.

The total cost of a product is made up of two types of costs known as **fixed costs** and **variable costs**. These are sometimes referred to as **overheads** and **prime costs**.

Fixed costs

Fixed costs occur all the time, whether production is taking place or not and include:

○ rent and insurance
○ monthly salaries
○ any business registration fees
○ water rates, phone line rental, etc.
○ depreciation of the cost of equipment.

Variable costs

Variable costs relate to the direct costs of production and therefore increase as production rises. They include:

○ raw materials
○ packaging and labels
○ labour
○ fuel
○ distribution and marketing
○ quality assurance.

When the total of the fixed costs and variable costs is equal to the sales value, the enterprise is said to be in a **break-even** position; covering all costs but making neither a profit nor a loss. As the production rate increases above the break-even mark, the business begins to make a profit.

Depreciation

It is normal practice to build an element into the price of the product that allows for the replacement of equipment when it reaches the end of its useful life. This is known as depreciation. There are several ways of calculating depreciation. The simplest is straight-line depreciation, i.e. 20 per cent loss in value each year over five years. Sometimes it is necessary to use a declining rate and residual value depreciation. For example, a vehicle costing $10 000 may lose $3000 in the first year, $2000 in the second and $1000 in the third. At this time it could be sold, having a residual value of $4000. The example shown below, using straight-line depreciation, has been found to be simple and understood by owners of small enterprises.

Step 1: Decide on the number of working days per year. Let us assume this to be 200.

Step 2: List all the equipment, with its cost, in groups based on an estimate of their useful life. In this example:

Step 3: Divide the cost of each group by the total

Group	Item	Life (years)	Cost ($)
1.	Building	20	10 000
2.	Major equipment: dryer, slicing machine	5	3000
3.	Minor equipment: heat sealer, scales, bin dryer	2	1000
4.	Consumables: knives, chopping boards, etc.	1	200

number of working days in the life to calculate a daily depreciation rate:

All the estimated fixed (including depreciation) and

Group	Life (years)	Cost ($)/No. working days	Daily depreciation ($/day)
1	20	10 000/(200 × 20)	2.5
2	5	3000/(200 × 5)	3.0
3	2	1000/(200 × 2)	2.5
4	1	200/(200 × 1)	1.0

Total depreciation = $9.0/day

variable costs (based on the planned output) are listed and added together. This gives the total costs per day. Let us assume that in this example the total cost per day is $100 and the planned output is 600 packets of dry food per day. From these figures, the unit cost to produce one packet can be calculated as:

$100/600 = $0.16 per pack

This is only a break-even price and does not give any profit. To calculate the wholesale price, a margin for profit (e.g. 25 per cent) is added to the break-even price, giving a figure of $0.20. The retailer also needs to make a profit of 25 per cent, therefore the final selling price is calculated at $0.25. This final price must be compared to the market survey that indicates the prices customers will be prepared to pay and the price of any competition.

6.4 Quality control and quality assurance

Food producers, however small, need to be aware that product quality is an area of increasing concern to consumers. If customers are not satisfied, in any way, with a product, they will try and find a better, alternative supplier. What is actually meant by the term quality? A generally accepted definition is that 'the product meets agreed customer expectations or specifications'. Too often small producers regard quality control as an afterthought or extra when it is really an important management tool that should be planned and costed into the production process.

It is important to appreciate the difference between quality control (QC) and quality assurance (QA). In the case of a dried food, quality control would involve checks during processing such as moisture content, raw material quality and net weights. QC therefore concentrates on checks during production. However, quality problems can occur after the product has left the production unit and these will reflect badly on the manufacturer even if they were not directly responsible for causing them. Examples of problems and how to tackle them are given in Table 6.1.

This wider approach, which follows the food through the chain to the point of consumption is known as **Total Quality Assurance**. Its focus is on prevention rather than cure.

In recent years, a quality management tool known as the **Hazard Analysis and Critical Control Point (HACCP)** system has become increasingly accepted. HACCP considers all the risks that a product may face that result in quality problems. These include raw materials, staff, the plant environment and the distribution chain. HACCP is a system that documents risks and identifies points at which monitoring is required to

Table 6.1 Examples of problems arising after a product has left the manufacturer

Problem	Course of action
A shopkeeper places the product in a window in strong sunlight, causing colour loss.	Maintain close contact with shops and carry out periodic visits and inspections.
Complaints of crushed packets due to the person transporting goods stacking other materials on top of them.	Discuss with transport firm.
Some packets have gone mouldy due to shop selling old and out of date stock.	Establish a system to buy back old stock, use clearer 'sell-by' labels and assist shop in maintaining better stock control.

minimize quality problems. These are identified critical control points. While HACCP is at present mainly used by larger companies, the principles can readily be applied by small producers with great benefit. Some buyers now insist that suppliers have a HACCP system in place to assure product quality.

Typical QC checks and QA measures

This section will briefly examine both QC and QA measures that can be used to maintain high quality in dried foods. It is strongly recommended that relevant literature in the bibliography is consulted to supplement this chapter.

Moisture content
The determination of the moisture content of a dried food is of great importance in order to ensure that it has been adequately dried. Moisture is determined by drying a small sample of the product for several hours in an oven at 105°C. Ideally, producers should be able to test their products on a routine basis. In reality, the high cost of the equipment required, in particular that of a very sensitive balance, means that in-house testing will be unaffordable.

Alternatively moisture can be determined using an infrared moisture tester. A sample of food is dried under an infrared lamp and the moisture content read off from a scale. These testers, while less accurate than the oven method, are less costly and give results in 7–10 minutes.

If neither of the above methods is appropriate, alternatives should be sought. These include:

o Sending samples at regular intervals to a university or institute with the required analytical equipment. This is essential if supplying a buyer with a specification for moisture.
o Establishing a system in which trays, which are clearly marked with their own weight, are loaded with a standard weight of raw material. After drying the trays are weighed and the weight of dry food is recorded. This check will ensure that each tray has been dried to the same degree.
o With experience, workers are often able to determine, to a surprisingly high degree of accuracy, whether a

food has been sufficiently dried by checking the crispness, the way it breaks or tears or the feel in the mouth.

○ When developing a new product, samples should be taken from the dryer at regular intervals and packed. Inspection of these dried samples at weekly intervals will show which develop, or do not develop, mould growth. In this way a standard drying time which has reduced the moisture content to a stable level can be determined.

○ It is recommended that daily shelf samples of the product are kept for the declared shelf life plus 25 per cent. Any sign of deterioration by mould growth after this time will indicate that the food was not sufficiently dry before packing or that the packaging is not providing sufficient protection against moisture pick-up from the atmosphere.

Sulphur dioxide levels

Checking sulphur dioxide levels is of particular importance when meeting a buyer's specification. The equipment and chemicals required are not expensive but operators require special training. It is suggested that a local institution is consulted if such testing is required.

Microbiological tests

It is likely that only large producers will have the facilities and skills required to carry out microbiological testing. It is, however, recommended that samples should be sent periodically for testing. Such testing will be essential if selling to a buyer with microbiological specifications.

Net weight

In almost every country the net weight of the product has, by law, to be clearly stated on packaged foods. Net weight checking should therefore be a routine step in the production system. It can also increase profitability by reducing the quantity of product that is given away unnecessarily. The balance used must have the required sensitivity and should be clearly marked with the net weight and the maximum permitted over-weight.

Hygiene

Worker hygiene is important if quality is to be maintained. Dirty, untrained and sick workers pose a real danger to food safety. The management process should assure that:

o Workers are trained in good hygiene practice. Remember people always perform better if they understand their importance in the process. Posters, such as that shown in Figure 6.3, help raise awareness.
o Toilets, a changing room and, if possible, a shower should be provided with **at least two doors** between them and the processing room.
o Workers are provided with clean white clothing and hats.
o Sick workers with stomach problems, coughs, septic cuts, etc., do not handle food.
o Individual workers have defined roles for cleaning and other hygiene duties.

WASH ALL CONTAINERS & STORE IN A CLEAN PLACE.

Figure 6.3 Good hygiene poster (© ITDG).

6.5 Worker health and safety

It is very important to consider the safety of workers in the processing plant. Severe injury caused by faulty management control may result in a worker taking legal action. The following areas should be considered:

o Electric outlets should be placed high on the wall where they will not be splashed.
o Electric cables should not run across floors, particularly if the floor is wet.
o All revolving machinery should be protected by guards.
o Fluorescent lighting should not be used where there is revolving equipment as this can appear to be stationary at certain speeds.
o Safety footwear and gloves should be provided. Workers should remove jewellery, ties or clothing that might catch in equipment.
o A first aid box should be readily available and one staff member should be given basic first aid training.

Further guidance on worker health and safety is to be found in the bibliography.

6.6 The business plan

If finance is required from a bank or lending institution, a business plan will have be prepared. This must demonstrate that all aspects of the proposed business have been investigated and considered and that the applicant has, or can obtain, all the required skills to run the business. A business plan for a small enterprise does not have to be complex and most lending bodies have literature to assist in its preparation. Basic data however, is essential for its preparation. As has been seen:

○ Market surveys will provide information about the potential amount of production and the type of packaging, presentation and distribution required.
○ The type and cost of the processing facility and the equipment required will be known.
○ Aspects such as marketing, quality control, raw material, labour, packaging costs and distribution will have been considered and costed.

A business plan describes what the business will do, who will buy its products and its costs. It also gives financial forecasts over at least one year or, more commonly, over the life of the loan requested. The preparation of the plan should not be seen in a negative way, but rather as a mental process that allows all aspects of the business to be investigated; it will form a road map for the development of the enterprise. Remember that those considering the plan are busy people. It is therefore important to be short and factual and provide the person reading the plan with a clear 'snapshot' of the business. There is no one way to prepare a business plan. The following information typifies the steps involved (courtesy of the Prince's Trust, which supports young entrepreneurs in the UK).

Content of the business plan

The following areas may need to be included in the plan. Some areas may not be appropriate, in which case they can be left out.

Summary
Provide a summary of the business and its aims and objectives. It is often best to write this section last in order to draw on the whole plan.

The business

Provide:
o name, address, telephone number and bank account of the business
o indicate the legal status of the business
o in the case of an existing business provide copies of records, accounts, bank records, etc.
o describe the state of the business to date – in the case of new enterprises explain what research has been done.

The product and service
o Describe your product/s. What need does it fill, what makes it needed?
o How much competition there is and who? Are they big or small companies? Have you strategies in place to deal with competition?

Market aspects
o What is the estimated size of the market, how is this likely to grow? On what basis has this demand been based? Are any market surveys available?
o Which customers are you targeting? Compare your products and prices with any competition – why are yours distinctive? What market share is expected for the business?

Marketing strategies
o Describe the marketing plan of the business. Is there a budget for marketing?
o Indicate the longer term business plan.
o Indicate any interest from potential customers.

Operational aspects
o The premises and equipment required should be described, together with a short description of the production process and plant capacity.
o List staff that will be required and their cost. Will they need training?
o In developing countries, specialist ingredients and packaging may not be locally available. How will such inputs be obtained?
o If imported equipment will be used, how will spares be obtained?

Management and organization

o Describe any specialist skills and knowledge that you and any associates have.

o Are there any weaknesses – if so how will they be overcome?

o Provide a short CV (curriculum vitae) with age, education and experience of all key workers.

Financial information

o Provide information on how the product has been priced.

o Give financial forecasts including cashflow, profit and loss and balance sheet projections. Calculate the break-even point at which the business will make a profit.

o If personal money is to be invested in the business, state how much.

o Detail any current financial commitments.

Financial requirements

o How much money is required and over what period?

o Will an overdraft be needed? If so, how much?

o List any grants or loans obtained or requested.

7 Guidelines for the design and construction of small mechanical dryers

IN MANY SITUATIONS mechanical dryers can be designed and built locally at costs considerably less than those of imported models. To do this the food producer has to work closely with engineers to ensure that the final design meets all requirements. This chapter examines the basic calculations required to design a system to dry a given amount of a particular food and aspects related to correct design for use when processing foods.

Experiences in several countries have shown that engineers can readily understand the steps involved in designing a mechanical dryer and can make modifications to overcome local constraints. However, it is strongly recommended that the food producer maintains close contact with the engineer, who may not have knowledge of good design in terms of food safety and hygienic design.

The design of a dryer involves the following basic steps:

1. Defining the quantity of food that is to be dried in one batch and the drying temperature, which normally lies between 55 and 65°C.
2. Measuring or obtaining data on the local climatic conditions (average temperature and RH) during the period of the year that the food will be dried.
3. Calculating the total amount of water that has to be removed from the food during the drying process.
4. Calculating the weight of moisture that 1 kg of air can remove under ideal conditions; i.e. assuming perfect transfer of moisture to the air.
5. As perfect transfer of moisture is never possible it is necessary to use what is commonly known

as the **pick-up factor**. This takes into account the nature of the particular food being dried and the ease with which it releases moisture to the air and allows the determination of the likely *actual amount* of water that each kilogram of air will remove. This then allows a calculation of the total weight of air required to remove the total amount of moisture.

6. Converting the weight of air, as found in step 5, to volume in order to size the fan (fans are normally specified by air volume).
7. Calculating the amount of energy that will be required to heat this volume of air from ambient to the chosen drying temperature.
8. The design of the whole system including the cabinet or drying chamber.

Steps 4 to 7 involve the use of the psychrometric chart, which is now described in greater detail.

7.1 The psychrometric chart

Psychrometry is the study of the properties of air under varying temperature and moisture conditions and therefore is of interest to drying technologists. The psychrometric chart was developed by heating and ventilation engineers to enable them to design air conditioning plants for buildings. Initially the relationship between air conditioning and drying may not be obvious but in fact there are aspects common to both. Air conditioning plant is required to maintain a suitable environment within an enclosed space, which may be an office block or a car or a building for storing inflammable chemicals. Whatever the needs, there are four functions required of an air conditioning plant:

○ heating air
○ cooling air
○ addition of moisture to the air (humidification)
○ removal of moisture from the air (de-humidification).

Two of these functions are required for drying, i.e. heating air and the removal of moisture from the food being dried to the passing air stream. **The psychrometric chart thus provides a useful tool to quickly calculate the**

heating and moisture removal requirements for any drying system, including foods.

On initial inspection the chart (see Figure 7.1) looks very complicated. However, with practice it is relatively easy to use and is quicker and simpler than calculating values from first principles. It consists of a series of lines that represent different properties of air under varying conditions.

The following terms used in the chart are defined as:

Dry bulb temperature The dry bulb temperature is the temperature of the air as measured by a standard thermometer.

Wet bulb temperature The wet bulb temperature is the temperature of the air when fully saturated. It can be found by enclosing the bulb of a standard thermometer within a wet cotton sock, which has the effect of simulating atmospheric conditions of 100 per cent relative humidity (RH). Unless the ambient conditions are 100 per cent RH the wet bulb temperature will always be less than the dry bulb temperature.

Percentage saturation lines The percentage saturation lines relate to the humidity of the air as a percentage of the absolute humidity of air that is fully saturated (i.e. no longer capable of holding moisture). They are in fact very similar to relative humidity and the values are often interchanged.

Adiabatic cooling lines The adiabatic cooling lines indicate what happens to the temperature and humidity

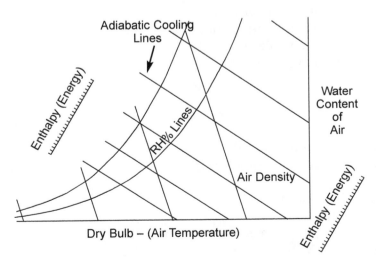

Figure 7.1 Simplified psychrometric chart.

of air when it is used for drying. The warm air provides the heat required for the evaporation of moisture from the food. The air will thus cool as moisture is evaporated from the moist food and absorbed by the warm air as it provides the heat required for evaporation. This reduces the capacity of the air to pick up more moisture and increases its relative humidity.

Specific enthalpy scales These relate to the energy contained in the air. The warmer the air the higher its energy or enthalpy. The scales are used to calculate the amount of fuel that will be required to heat the air used to dry the food.

Specific volume lines The volume of any gas, including air, will vary with temperature. These lines represent the change in volume for air at a given temperature.

Absolute humidity This is the weight of water in each unit weight of air. It is normally expressed as kg of water per kg of air.

7.2 Calculating the sizes of fans and heaters

This section will, by means of an example, follow steps 1 to 7, as listed in the introduction, to design a small mechanical tray dryer.

Define the amount of food to be dried and the ambient climatic conditions

In this example it is assumed the producer wishes to dry 200 kg of fresh herbs per day. The required drying time is ten hours to allow for loading and unloading of the dryer.

Obtain the local climatic conditions

Local average daily ambient conditions during the period of the year when the food will be dried, as supplied from local weather data, are 18°C and 60 per cent relative humidity (RH).

Calculate the weight of water that has to be removed per batch

The producer knows, from discussions with a local institution, that the moisture content of the fresh herb is approximately 90 per cent. The buyer of the final dry product has indicated that the maximum moisture should be 6 per cent.

The batch weight is 200 kg at 90 per cent moisture. Thus, total weight of water present is:

$$200 \text{ kg} \times (90/100)$$

i.e. 180 kg of water, which means that the dry matter (with zero per cent moisture) will be:

$$200 \text{ kg} - 180 \text{ kg} = 20 \text{ kg}$$

After drying the dry matter will still be 20 kg which will be equivalent to 94 per cent of its weight at 6 per cent moisture.

Thus at 6 per cent moisture content the final weight of the product will be:

$$20/0.94 = 21.28 \text{ kg; i.e. } 20 \text{ kg dry matter and } 1.28 \text{ kg}$$
$$\text{of water}$$

This means that the weight of water that has to be removed is:

$$180 \text{ kg} - 1.28 \text{ kg; or } 178.72 \text{ kg per batch of } 200 \text{ kg.}$$

Calculate the quantity of air required to remove the required moisture from the batch

It is now necessary to use the psychrometric chart as shown in Figure 7.2 using the agreed ambient conditions of 18°C and 60 per cent RH.

First determine the moisture content of the ambient air using the psychrometric chart shown in Figure 7.2 as follows:

○ Draw a line from the dry bulb temperature of 18°C vertically upward until it crosses the 60 per cent saturation, or relative humidity, line at point A.
○ Next draw a horizontal line from point A to the right hand vertical axis (point B) and read off the moisture content for this temperature, which is 0.0078 kg moisture per kg air.

Next, determine the maximum amount of moisture that this air can hold when it is heated to 50°C using the psychrometric chart (Figure 7.2) as follows:

○ Draw a vertical line from the selected drying temperature of 50°C to meet the horizontal line A–B at point

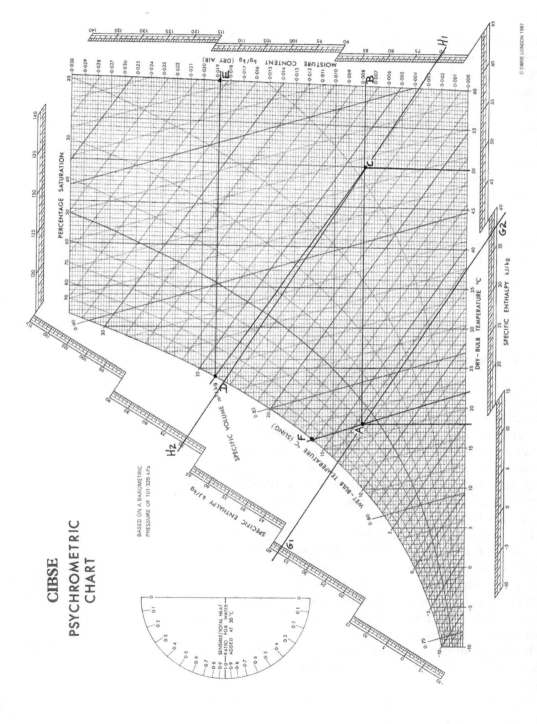

CIBSE
PSYCHROMETRIC CHART

BASED ON A BAROMETRIC
PRESSURE OF 101·325 kPa

© CIBSE LONDON 1987

Figure 7.2 Psychrometric chart.

C. It should be noted that the air, after heating, still contains the same amount of moisture but its RH has fallen from 60 per cent to approximately 10 per cent. In other words, it has a greater capacity to absorb moisture from the food.

o Now draw a line, from C, parallel to adiabatic cooling lines to meet the 100 per cent RH line at D. (This line can be considered as the result of passing the air through trays of food and so absorbing moisture until it becomes saturated.)

o Finally draw a horizontal line to the moisture content axis at point E and read off the moisture content which is:

0.0188 kg moisture per kg air

The theoretical weight of moisture that each kg of the heated air can remove can now be calculated:

moisture content at E – moisture content B
= 0.0188 – 0.0078

= 0.011 kg moisture per kg air

Use of the pick-up factor

It is the physical and chemical composition of a food which to a very great extent controls the rate at which it will dry (see Chapter 2). Fibrous materials that are high in cellulose lose moisture easily, while sweet fruits dry much more slowly due to the fact that the sugar binds moisture. The pick-up factor, which is largely empirical and determined by drying trials, attempts to adjust drying calculations by taking into account the nature of the particular food to be dried.

It has been found that pick-up factors range from 0.15 to 0.40, that is to say that between 15 per cent and 40 per cent of the theoretical moisture pick-up of the heated air, as calculated above, will in fact occur. Leafy materials have high pick-up factors around 0.40, starchy foods have values of approximately 0.25 and high sugar and salt food are as low as 0.15.

In this example, a herb is being dried and a pick-up factor of 0.40 (or 40 per cent) is a reasonable assumption.

Thus, the likely actual loss of moisture to the air will be:

0.011 × 0.4 = 0.0044 kg moisture per kg of air

In Step 3 it was calculated that 178.72 kg of moisture need to be removed during the ten hour drying cycle to reduce the moisture content of the food to 6 per cent.

Thus, the total number kilograms of air required is:

$$178.72/0.0044 = 40\ 618 \text{ kg of air}$$

Convert the weight of air to volume of air

Blowers or fans are usually specified by the volume of air they can deliver per hour which is known as the volumetric flow rate and is measured in cubic metres of air per hour, m³/h, or cubic feet per minute, cfm (1 cfm = 1.699 m³/h and 1 m³/h = 0.588 cfm). If the total amount of air needed and the time required to dry the product is known then, with the help of the psychrometric chart, the volumetric flow rate can be determined as follows:

volumetric flow rate (m³/h) = mass flow rate (kg/h)
× specific volume (m³/kg)

Where the mass flow rate per hour is equal to the total amount of air needed (kg) divided by the time required to dry the product. In this example airflow rate is:

mass flow rate (M_a) = 40 618 kg/10 h = 4061.8 kg/h

The specific volume can be found using the chart by drawing a line through point A (18°C at 60 per cent RH) parallel to the specific volume lines and reading off the specific volume at point F. The reading in this case falls approximately half way between the 0.83 and 0.84 line and so can be taken 0.835 m³/kg. Thus:

volumetric flow rate = 4061.8 kg/h × 0.835 m³/kg
= 3391.6 m³/h

Calculate the amount of energy required to heat this volume of air to the chosen drying temperature

The energy (or heat) required is equal to the weight of air required per second multiplied by the difference between the specific enthalpy (equivalent to energy it contains) of the air before and after heating. This is shown in the following formula:

$$Q = M_a \times (h_2 - h_1)$$

where: Q = heat required (kW); M_a = air mass flow rate (kg/s); h_1 = specific enthalpy of air at inlet to the heater; and h_2 = specific enthalpy of air at drying temperature used.

The mass airflow rate has already been calculated using the psychrometric chart (see previous step) although it is given in kg/h. This value will have to be divided by 3600 to convert it to kg/s.

Thus:

$$(4061.8 \text{ kg/h})/3600 = 1.128 \text{ kg air/s}$$

The psychrometric chart is now used to find the specific enthalpy for the air before and after heating (i.e. at 18°C and 50°C).

There are two specific enthalpy scales marked on the chart (see Figure 7.2) that have to be used together, using a rule long enough to span the chart.

Using point A as a pivot, adjust the rule so that identical values can be read from the two scales as shown by the line G_1–G_2. In this case the reading should be 37.5 kJ/kg.

Repeat using point C as the pivot taking the reading at H_1 and H_2. In this case the reading should be 70 kJ/kg.

The amount of heat can now be calculated using the formula:

$$Q = M_a \times (h_2 - h_1)$$

Therefore:

$$Q = 1.128 \text{ kg/s} \times (70 - 37.5 \text{ kJ/kg}) = 36.7 \text{ kJ/s}$$

Since 1 kJ/s = 1 kW, the heat required is 36.7 kW (although all burners are less than 100 per cent efficient and this is discussed in Section 7.7).

In the worked example above it is assumed that the food manufacturer and engineer are designing a dryer from first principles with no equipment available. In reality this is usually not the case. For example, the producer may have a fan or a heater and require a dryer to be built around these components. The psychrometric chart can be used to solve all such situations. The following case study demonstrates another way the chart can be used.

7.3 Fish drying case study

A project was implemented to develop the markets for dried fish produced by small enterprises on a South

Pacific island. The project requirements were to dry 100 kg of fish in six hours using simple tray dryers. Heat for the dryer was provided by a simple biomass burner fuelled by wood. Before drying, the fish was brine-treated to reduce initial moisture content and suppress the growth of micro-organisms.

The following conditions applied:

o average ambient daytime air conditions: 28°C and 80 per cent relative humidity
o moisture content after brining: 60 per cent (wet basis)
o final moisture content of fish should be: 25 per cent (wet basis)
o maximum drying temperature: 40°C to avoid case hardening.

The psychrometric chart (Figure 7.3) was used to calculate the amount of fuel required for drying, the size of fan required, the drying time, and allowed for calculation of production costs to be made.

Calculation of the weight of moisture to be removed

The brined fish has a moisture content of 60 per cent so 100 kg of brined fish will contain 60 kg of moisture and 40 kg of dry matter. After drying the moisture content will be reduced to 25 per cent and the dry matter will still be 40 kg. Thus the total weight of the finished product at 25 per cent moisture will be:

$$40 \text{ kg}/0.75 = 53.3 \text{ kg}$$

The weight of moisture present in the final product = final weight minus the dry matter weight; i.e.

$$53.3 - 40 \text{ kg} = 13.3 \text{ kg}$$

Therefore the total weight of moisture to be removed during drying is the initial moisture content (60 kg) minus the final moisture content (13.3 kg):

$$60 - 13.3 \text{ kg} = 46.7 \text{ kg}$$

Calculation of the quantity of air required to remove this weight of water

Using the chart, draw a line vertically upward from 28°C on the dry bulb temperature line until it crosses the

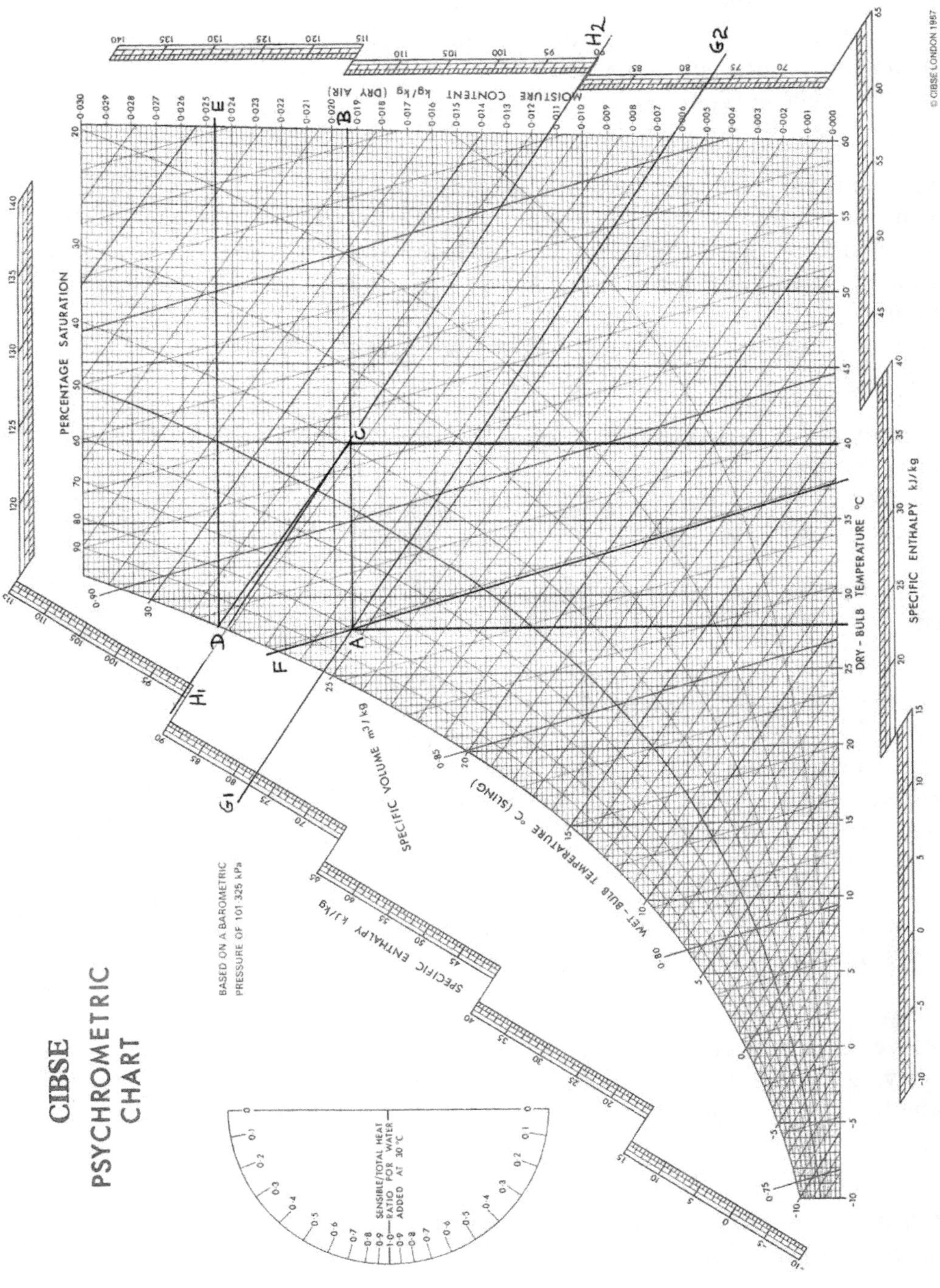

Figure 7.3 Psychrometric chart for fish drying case study.

80 per cent relative humidity line (point A). Read off the moisture content from the vertical axis (point B), which gives 0.0194 kg moisture per kg air.

Next, calculate the theoretical weight of moisture that the heated air can hold. Draw a vertical line from the 40°C point to where it crosses the horizontal line A–B at point C. From this point draw a line parallel with the adiabatic cooling line until it crosses the equilibrium saturation line at 100 per cent RH (point D). Read the moisture content on the vertical axis at point E, which gives 0.0248 kg moisture per kg air.

The potential for heated air to pick up moisture is the moisture content at point E minus the moisture content at point B; i.e.

$$0.0248 - 0.0194 = 0.0054 \text{ kg moisture per kg air}$$

This is the theoretical amount of water that the air can remove and the pick-up factor now has to be used. For fish drying a pick-up factor of 0.2 was assumed reasonable. Thus, the actual weight of moisture likely to be removed per kg of air is:

$$0.0054 \times 0.2 = 0.0011 \text{ kg moisture per kg air}$$

The total quantity of air required to dry the 100 kg batch of fish is the total amount of moisture to be removed divided by the actual potential for air to pick up moisture, thus:

$$46.7 \text{ kg}/0.0011 \text{ kg moisture per kg air } = 42\,454 \text{ kg of}$$
$$\text{drying air over six hours}$$

or 7075 kg/h (this is equivalent to the mass flow rate).

Calculation of required fan size

The mass flow rate, calculated above, now has to be converted to a volumetric flow rate in cubic metres.

The chart is now used to find the specific volume of the air. Draw a line through point A parallel with the specific volume lines. The reading (point F) should be 0.879 cubic metres per kg.

Thus the volumetric flow rate is:

$$7075 \times 0.879 = 6296 \text{ m}^3/\text{h}$$

At this point it became apparent that a very large fan would be required to dry the fish in six hours.

Discussions with the producers produced the following options:

○ reduce the batch size
○ extend the drying time
○ increase the drying temperature.

It was agreed that the drying time could be extended to ten hours without lowering product quality. Thus, the mass flow per hour with a ten-hour drying time is:

$$42\ 454/10 = 4245 \text{ kg of air per hour}$$

The revised volumetric flow rate is thus:

$$4254 \times 0.879 = 3739 \text{ m}^3/\text{h}$$

Fans that could supply this quantity of air were locally available.

Estimating the energy required from burning wood to heat this quantity of air

The heat required to dry the product is found using the following formula:

$$Q = M_a \times (h_2 - h_1)$$

where: Q = heat required (kW); M_a = air mass flow rate (kg/sec); h_1 = specific enthalpy of air at inlet; and h_2 = specific enthalpy of air at drying temperature.

To find the specific enthalpy of air at the inlet, lay a rule through ambient conditions (28°C, 80 per cent RH) and, using point A as a pivot, adjust the rule until the same reading is achieved on both enthalpy scales (line G_1–G_2) to give the enthalpy of inlet air as 77.5 kJ per kg air.

Repeat this procedure to find the specific enthalpy of the heated air (40°C, 80 per cent RH) using point C as the pivot and adjust until the same reading is achieved on both enthalpy scales (line H_1–H_2), giving the enthalpy of heated air as 90.5 kJ per kg air.

Therefore:

$$Q = 1.179 \text{ kg/s} \times (90.5 - 77.5 \text{ kJ/kg}) = 15.33 \text{ kJ/s}$$
$$\text{or } 15.33 \text{ kW}$$

Assuming a calorific value of 15 000 kJ per kg for dry wood (adjusted for the expected burning efficiency), then:

15.33 kJ per second/15 000 kJ per kg
     = 0.0010 kg per second

This is equivalent to burning 3.68 kg wood per hour or approximately 37 kg of wood per ten-hour drying cycle.

7.4 Design and construction of dryer cabinets

The aim of this section is to provide general guidelines for consideration when undertaking the design of drying chambers such as cabinets and bins. It does not pretend to offer specific information on the construction of any particular dryer.

The high capital cost (US$30 000 or more) of small dryers from high income countries, such as the United States of America and Japan, makes them uneconomical for small business in developing countries. Dryers of this type are almost entirely used in test or research facilities. Adequate dryers for small enterprises in developing countries can be locally constructed at reasonable cost, using local materials and skills.

It is often said that good design is 10 per cent inspiration and 90 per cent perspiration; in other words the main problem is putting a good idea into practice. Getting the design right, or as right as possible, before starting to manufacture saves time and money and may represent the difference between profit and loss. It is important, therefore, that the designer has a good understanding of the customer's requirements. This involves a close interaction between the client and the constructor. The designer should be prepared to visit clients to examine, at first hand, any problems with equipment in order to incorporate changes and improvements in future designs.

There are three basic elements to the design process: function, construction and form.

Function

This element is perhaps the most important to the designer as it deals with how well the dryer performs in practice and meets customer requirements. Function not only takes account of what the dryer is being used for but the way in which it is used. It is difficult to get all functional aspects of design correct. Unless money is no object, the final design will be a compromise between performance and cost. There is no point in producing a dryer that is absolutely perfect if nobody can afford to buy it. Therefore the dryer will be designed and manufactured to a price that is acceptable to both the customer and manufacturer, which will mean some trade-off in performance against cost.

The designer will need to have or gain a good understanding of the practicalities of drying foods so that value judgements can be made as to areas in which compromises can be made.

Typical questions that the designer should ask include:

o What is the precise role of the dryer?
o To what accuracy must the drying be performed (i.e. drying temperature, time, final moisture)?
o What is the operating capacity of the dryer (e.g. 20 kg of fish per hour)?
o Under what conditions must the dryer operate (e.g. hot, cold, wet, inside, outside, etc.)?
o Are there any limitations regarding the method of operation (e.g. manual, electrical, etc.)?
o Will the dryer be used frequently?
o How reliable must the dryer be?
o Are there any safety requirements? For example, when drying powder products such as starch, which are prone to violent dust explosions, it would be necessary to ensure that all flames and sparks are isolated from the drying area.
o Are there any legal or insurance requirements that must be met?
o What is the maximum cost the client will pay for the equipment?

Construction

This element of the design deals with the way in which the separate components of the design fit together to form the finished dryer. A good, well-thought-through design will mean that many parts can be made before assembly begins, perhaps by different workshops, and when assembled will fit together.

Typical questions that should be asked include:

o How strong must the dryer be?

- How large will the dryer be?
- How complex is the dryer; will it require special construction techniques?
- Will the environment in which the dryer is used affect material choice (i.e. will corrosion, high temperatures, acidic conditions, etc. cause problems for the material)?
- Are there any aspects of assembly or maintenance that will affect construction?
- Are there any international, national or local regulations (e.g. safety, legal, insurance, etc.) that could affect the method of construction (e.g. boiler safety regulations dictate the welding methods required for pressure vessels)?
- Are there any construction limitations imposed by the availability of equipment and materials to the manufacture of the product?

Form

This is the aesthetic part of the design, in other words what the finished dryer will actually look like. Although the designer might think that function is more important than appearance, the customer may disagree. The customer will often have a preconception of what a particular type of dryer looks like and therefore will be expecting something similar. Aspects of the design such as finish, colour, and shape may be just as important to the customer as performance. It is useful to remember that many people, technical and non-technical, judge products on the principle that if they look right then they are right.

Typical questions that should be asked include:

- Is good appearance an important selling point?
- Is colour important as a means of identification?
- Does shape, texture or colour improve the product or extend its life?
- Does the method of manufacture affect the shape and appearance of the product?

To summarize, a good designer should be prepared to seek new solutions and be inventive. They must, however, be prepared to justify all aspects of the design in terms of function, construction and form.

7.5 Materials selection

Having considered with the client the aspects of design described above, the designer will be in a position to begin to identify construction materials and components, wherever possible from the local market. Appropriate material selection is very important. If expensive specially imported materials are chosen the final cost of the dryer may be too high. If poor quality materials are chosen the dryer may soon fail to operate properly or will be expensive to maintain.

While the ideal material for the drying chamber and tray construction is stainless steel, its cost is prohibitive, and cheaper, alternative materials could be considered. However, when considering alternatives it is important to assess their suitability in terms of strength, ease of construction and, most importantly, safety.

The following notes are a guide to alternative materials used in the design and construction of small-scale dryers in a number of countries:

o Stainless steel mesh for the trays can be expensive so food grade plastic mesh may be suitable and cheaper. However, it will need to be tested to ensure that it can cope with the heat without destruction or deforming. A simple test would be to place a sample of the plastic in hot water to see if it can withstand the heat.

o Untreated, dry and well-seasoned wood can be used to make the tray frames on to which the mesh is attached. Treated wood should not be used unless it can be ensured that no toxic chemicals are used in the treatment process. For example, wood treated to withstand termite and ant attack is often treated with a cyanide solution.

o Timber that splinters easily should not be used as splinters may contaminate the product and could cause injury to the consumer and also to those handling the trays.

o Plywood attached to a wood or metal frame makes a good option for cabinet construction but ensure that the adhesive used in the manufacture of the plywood is safe to use with food.

o Cabinets should only be painted externally with a food grade paint. Lead-based should never be used to paint any part of the dryer. Trays should never be painted.

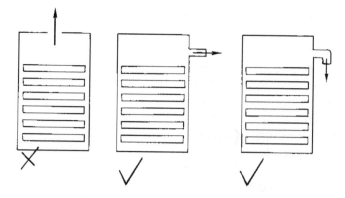

Figure 7.4 Two alternative air exit designs.

○ Steel can be used for the construction of the dryer in areas where it does not come into contact with food.

○ Galvanized steel can be used, but not for parts that come into contact with the food. The zinc coating can separate and contaminate food products.

○ Aluminium is light and cheap but does oxidize, forming a film on all exposed surfaces. This oxide film actually protects the aluminium from further oxidation and care should be taken not to polish or scrape this protective oxide film away.

7.6 Design considerations

There are two fundamental aspects to the design of any dryer. The first is concerned with the food; avoiding contamination, over-drying, under-drying, unequal drying, etc. The second is concerned with the safety and well-being of the operators.

Food safety considerations

Many aspects relating to food safety have been highlighted in the section on materials selection (Section 7.5). Additional points include:

○ The air exit at the top of the drying chamber should be designed in such a way that dust and other materials cannot fall into the dryer. This can be achieved by the use of an elbow-shaped duct or locating the exit at the top of one of the walls above the topmost tray (see Figure 7.4).

o It is advisable to cover the air exit duct with mesh to prevent insects and birds from entering the dryer when not in use.

o Care must be taken to prevent contamination by the products of combustion from the heat source. Ideally, when using combustion heaters (wood, gas, kerosene, diesel, etc.), indirect burners with a heat exchanger should be used, ensuring that all combustion products are exhausted to the atmosphere outside the building. This prevents food contamination and ensures that the operators do not breathe in fumes. This does not apply when using electric heaters. In reality, however, in many situations direct gas heaters are used as gas burns much more cleanly than solid or oil-based fuels as long as the burner is designed correctly and is well maintained.

o Product overheating can be caused by either poor operator control or by the use of thermostats with insufficient sensitivity. If possible, thermostats working to plus and minus 5°C or better should be used. Proper training of operators is required, particularly where wood-fired heating is involved.

Ergonomics and operator safety and well-being

To maintain high levels of productivity it is important that operators are confident that the dryer can be operated safely and this does not cause unnecessary strain or discomfort. The main task of dryer operators is the loading and unloading of trays of material. The trays, particularly when full of wet material, can be heavy. There-

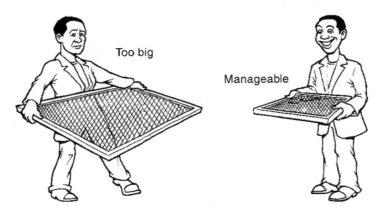

Too big

Manageable

Figure 7.5 Badly designed trays make the worker's job more difficult.

fore consideration must be given to how the dryer is operated:

- The tray design is important as the width and length of the tray will determine how many people will be required to load and unload from the dryer. If a single person is to load the trays, the width should not be such as to cause discomfort to the operator and the length should not be too great as to cause strain to the operator's back. Good and bad tray design is shown in Figure 7.5.
- The height of the cabinet should be determined by the maximum height that operators can comfortably reach while loading a tray, without becoming out of balance or the tray becoming too difficult to hold. If a tall cabinet is required then a safe and easy means of loading the higher trays needs to be devised.
- It should be possible to load and remove trays from the cabinet without risk of injury to the operators. Care is needed to ensure that operators' hands and fingers cannot become trapped between the trays or in the loading doors. Handles fitted to the front of the tray enable the operators to remove the trays without having to put their hands inside the cabinet.
- Where and how the trays are loaded should be considered. Loading on the floor means that operators need to crouch or bend for long periods, which may cause fatigue. Also, lifting a heavy tray from the floor, so that it can be placed in the dryer, may also result in operator fatigue or perhaps injury. Using a table or platform on which the trays can be loaded will enable the operators to perform their task more comfortably than bending or crouching.
- All fixings such as door handles and locks need to be sufficiently large to enable operators to hold them securely and robust enough to prevent them being damaged under normal operating conditions. Tray support rails should be robust enough to withstand fully loaded trays being dropped onto them.
- Safety features such as cut-off switches and emergency stop buttons should be large and placed so that under any operating condition they are easily accessible by the operators.
- All controls that are used in normal operation should be large enough to be operated with a gloved hand,

robust enough to withstand regular use, and easily accessible during all operation conditions.

7.7 Heat sources

Types of heaters

A range of heat sources may be used for drying food on a small scale, including solar, biomass, gas, oil and electricity. The choice will depend on the cost and availability of a heat source, the local climate and its appropriateness to the application.

At its most basic, solar heat is free, requiring the food to be laid out in the sun and dried over a number of hours or days. More sophisticated designs include the use of plastic or glass sheeting to cover the food product and some means of controlling the flow of air over the product. In all cases the heat available for drying is dependent on ambient temperatures, local weather patterns and the number of daylight hours unless it is possible to store heat produced during the day.

Electric heaters, similar to those used for cooking stoves, can be used as the heat source, but they need to be of good quality as they will be used for hours at a time. However, electricity can be expensive and a reliable supply is required. Failure of the electricity supply could result in spoilage or loss of the complete batch.

Biomass furnaces can be a cheap and effective means of providing drying heat regardless of the time of day, ambient conditions and local weather patterns. Biomass such as wood, wood waste, charcoal, coconut husk and rice husk can all be used as fuel, although the furnace or burner usually has to be designed to operate on one particular type of fuel. Biomass fuels are solid and do not burn as well as a gas and so could contaminate the food being dried, therefore they should only be used indirectly with a heat exchanger.

Gas burners, if designed properly and well maintained, will burn cleanly and give long reliable service. They can be used directly or indirectly to dry the food product. If they are being used directly then it is very important that the burner is well maintained so that the products of combustion cannot contaminate the food. Also, care is needed to ensure that the flame cannot come into contact with the food, causing it to cook or to burn. Direct-fired gas burners should never be used when drying fine powders, such as starch, as there is a high risk of a dust explosion. It should be remembered that the emissions from burning biomass or fuel oil can be harmful to health.

Burner efficiency

Whatever the source of heat used there will be some efficiency losses from the conversion of fuel to heat. Typically:

○ Electric heaters are almost 100 per cent efficient.
○ Direct gas burners are 80 to 95 per cent efficient.
○ Oil fired burners are 75 to 85 per cent efficient.
○ Solid fuel (coal, wood, agricultural waste) burners are 50 to 60 per cent efficient.

Standard heaters are sold with a rated heat output, which takes into account the conversion efficiency. As long as they are regularly maintained and correctly operated they should maintain this efficiency over many years. When designing burners and heaters, however, it will be necessary to take into account the expected efficiency to ensure that the required heat output is achieved.

Heat exchangers

When using a burner or heat source that could contaminate the food, a heat exchanger will be required to transfer heat from the heat source to a fluid such as air or water. Typically, the type of heat exchangers used to heat air for drying purposes are the finned tube or coil type with the heat passing through the tube or coil and air being blown across the outer surface.

The design theory of heat exchangers is complex and beyond the scope of this book. However, some simple heat exchangers can be made. Probably the most common heat exchanger to be found is a truck or car radiator, which is a finned tube type. Hot water circulates through the radiator and air is then blown through the radiator into the dryer. Most types of vehicular radiators also have dedicated fans or blowers fitted to them, providing a combined heat/fan set. One of the main advantages of using a water based heat exchanger such as a vehicle radiator is that the water can be heated some distance away and then circulated through the radiator.

This minimizes the possibility of contamination of the food by the fuel being burnt.

Fans and blowers

The are two basic types of fan: axial and centrifugal. Axial fans are similar to those seen on cars or air extractors, while centrifugal fans consist of a rotor in a housing with the air entering through the side of the housing and leaving via an outlet at right angles to the inlet. Centrifugal fans provide a higher pressure and they can thus drive air through deeper beds of material, against the back-pressure.

The calculations carried out earlier in this chapter will have defined the required airflow rate and, hence, fan size. From the designer's point of view there are two main considerations:

○ The fan must have enough power to overcome the back-pressure caused by the trays of food.
○ The airflow through the trays must not be too high as to cause the food particles to be blown off the trays; remembering that the dry material will be lighter than the fresh.

Excessive back-pressure may be overcome by use of a more powerful fan, reducing the depth of food on each tray and generating a zigzag airflow through the cabinet as described earlier.

Reducing the number of trays and increasing their area will reduce the linear airflow through the trays and overcome tendencies for particles of food to blow off the trays.

Food drying is not an exact science and practical experience is as important as theory. It is hoped that the guidance notes in this chapter will give engineers and food processors the confidence to develop appropriate solutions to their drying problems.

Long experience of designing and developing dryers for different applications has taught the authors that it is a stepwise process. There is no better way to learn than by doing.

Glossary

Absolute humidity	the weight of water in a given weight of air.
Aflatoxins	cancer producing poisons produced by moulds of the Aspergillus group that are found in poorly dried cereals, oilseeds and nuts.
Blanching	the heat treatment of food by hot water or steam in order to inactivate enzymes.
Brix	a scale used to measure sugar levels in a food, measured by a refractometer.
Case hardening	formation of a dry skin on the surface of a food if it is dried too quickly. Drying rates are reduced.
Chlorination	Addition of chlorine to water to destroy micro-organisms.
Enzymes	natural food proteins that may cause changes to colour, flavour and texture.
Equilibrium relative humidity	the moisture content at which a food does not gain or lose moisture to air.
Ergonomics	design of equipment that takes into account safe and comfortable use.
HACCP	Hazard Analysis and Critical Control Point system, used to control contamination in food processing.
High acid foods	foods with a pH below 4.5 that cannot support the growth of food poisoning micro-organisms.
Hygroscopic	substances that readily absorb moisture from the atmosphere.
Low acid foods	foods with a pH above 4.5 that can support the growth of food poisoning micro-organisms.
Lye peeling	removal of skin by hot caustic soda solution.
Micro-organisms	minute forms of life, including moulds, yeasts and bacteria.

Net weight	the weight of food filled into a container.
Osmosis	a process in which water is removed from a food by immersion in a strong solution of salt or sugar.
ppm	parts per million.
Psychrometric chart	a graphical representation of the properties of air commonly used when undertaking drying calculations.
Quality assurance	management systems which assure that food is safe and in a condition expected by the consumer.
Quality control	checks and tests that ensure that a food has the required quality specification.
Refractometer	an instrument used to measure soluble solids, such as sugar in jams or salt in brines.
Sodium metabisulphite	chemical preservative effective against yeasts.
Sulphiting	treatment of a food with sulphur dioxide in the form of sodium bisulphite or sodium metabisulphite.
Sulphuring	treatment of a food with sulphur dioxide gas.

References and further reading

References

Begum, S. (1986) *Small-scale Food Dryers* (literature review concentrating on Peru, Bangladesh, Kenya and Sri Lanka), ITDG, Rugby.

Brenndorfer, B., Kennedy, L., Oswin Bateman, C.O., Trim, D.S., Mrema, G.C., and Wereko Brobby, C. (1987) *Solar Dryers – Their Role in Post-Harvest Processing.* Commonwealth Science Council, Commonwealth Secretariat Publications, London (ISBN 0 85092 282 8).

Doe, P.E. (1979) *A Polythene Tent Fish Dryer – A Progress Report.* Proceedings of an international conference 'Agricultural Engineering in National Development', University of Pertanian, Selangor, Malaysia.

Hilario, R. (1999) Unpublished report on use of test dryer, ITDG–Peru.

International Trade Centre (1994) *Dried/Dehydrated Tropical Fruit: A Survey of Major Markets.* UNCTAD/GATT.

Lozada, E.P. (1983) *The Los Banos Multi-Crop Dryer.* University of the Philippines, Laguna, Philippines.

MAFF (1995) *Miscellaneous Food Additives Regulations.* Her Majesty's Stationery Office, 49 High Holborn, London.

McCance, R. and Widowson, E. (1991) *Composition of Foods*, 5th Edition. Royal Society of Chemistry/Her Majesty's Stationery Office, 49 High Holborn, London.

McDowell, J. (1973) *Solar Drying of Crops and Foods in Humid Tropical Climates*, Report CFNI-T-7-73, Caribbean Food and Nutrition Institute, Kingston, Jamaica.

NRI (1998) *Table of Recommended Thickness of Various Fruits for Drying.* Natural Resources Institute, Chatham, UK.

Ortiz, N.A., Cooke, R.D. and Quiros, M.R. (1982) 'The Processing of a Date-like Caramel from Cashew Apple' in *Tropical Science* 24(1): 29–38.

Platt, S. (1994) ITSL *Tray Dryer Field Trials.* Internal report, ITDG–Sri Lanka.

Salter, D. (1988) 'Coconut Processing in Vietnam', in *Food Chain* 22, ITDG, Rugby.

Useful further reading

Food science and composition

Egan, H., Kirk, R.S. and Sawyer, R. (1981) *Pearson's Chemical Analysis of Foods*. Churchill Livingstone, London (ISBN 0 443 02149X).

Paul, A.A. and Southgate, D.A.T. (1985) *The Composition of Foods*. Her Majesty's Stationery Office, 49 High Holborn, London.

Tan, S.P., Wenlock, R.W. and Buss, D.H. (1985) 'Immigrant Foods', 2nd Supplement to 4th Edition of McCance and Widdowson's *Composition of Foods*. Her Majesty's Stationery Office, 49 High Holborn, London (ISBN 0 11242 7170).

Drying

Axtell, B.L. and Bush, A. (1991) *Try Drying It. Case Studies in the Dissemination of Tray-Drying Technology*. ITDG Publications, London (ISBN 1 85339 039 9).

Brennan, J.G. (1999) *Food Dehydration: A Dictionary and Guide*. Woodhead Publishing, Cambridge, UK (ISBN 1 85573 360 9).

Brennan, J.G. (1994) *Food Dehydration*. Butterworth-Heineman, Oxford, UK (ISBN 0 7506 1130 8).

Brett, A., Cox, D.R.S., Trim, D.S., Simmons, R. and Anstee, G. (1996) *Producing Solar-Dried Fruit and Vegetables for Micro- and Small-Scale Rural Enterprise Development*. Natural Resources Institute, Chatham.

Esper, A. and Mühlbauer, W. (1996) 'Solar Tunnel Dryer for Fruits', in *Plant Research and Development* 44: 61–80.

Exell, R.H.B. and Kornsakoo, S. (1979) 'Solar Rice Dryer', *Sun World* 3(3): 75.

Greensmith, M. (1998) *Practical Dehydration*. Woodhead Publishing, Cambridge, UK (ISBN 1 85573 394 3).

Norman, G.A. and Corte, O.O. (1995) 'Dried Salteds Meats: charque and carne-a-sol,' FAO Animal Production Paper. FAO, Rome.

Oti-Boatang, P. and Axtell, B. (1993) *Drying*. UNIFEM/ITDG Publishing, London (ISBN 1 85339 308 8).

Rozis, J.F. (1997) *Drying Foodstuffs*. Backhuys Publications, Leiden, The Netherlands (ISBN 90 73348 75 7).

Van Arsdel, W.B., Copley, M.J., and Morgan, A.I. (1973) *Food Dehydration*, Vols 1 & 2, Ari Publishing, Westport, CT, USA.

Food processing and commodities

Adams, L. and Axtell, B.L. (1993) *Root Crop Processing.* UNIFEM/ITDG Publishing, London (ISBN 1 85339 138 7).

Asiedu, J.J. (1989) *Processing Tropical Crops.* MacMillan Press, London (ISBN 0 333 44857 X).

Fellows, P.J. (1993) *Food Processing Technology* (2nd Edition, 2000). Woodhead Publishing, Cambridge, UK.

Fellows, P.J. and Hampton, A. (1992) *Small Scale Food Processing: A Guide to Appropriate Equipment.* CTA/ITDG Publishing, London (ISBN 1 85339 108 5).

Ihekoronye, A.I. and Ngoddy, P.O. (1985) *Integrated Food Science and Technology for the Tropics.* Macmillan Press, London.

UNIFEM *Fish Processing.* UNIFEM/ITDG Publishing, London (ISBN 1 85339 137 9).

UNIFEM *Fruit and Vegetable Processing.* UNIFEM/ITDG Publishing, London (ISBN 1 85339 135 2).

Pursglove, J., Brown, C. and Robbins, S. (1981) *Spices* (Vols 1 & 2). Longmans, London.

Walker, K. (1995) *Practical Food Smoking.* Neil Wilson Publishing Ltd, Glasgow (ISBN 1 897784 45 7).

Packaging, marketing, sanitation and quality assurance

Anon. (1989) *Disinfection of Rural and Small Community Water Supplies.* Water Research Centre, Medmenham, UK.

Dillon, M. and Griffith, C. (1995) *How to HACCP.* MD Associates, Lincolnshire, UK (ISBN 1900134 00 4).

FAO (1969) *Codes of Hygienic Practice of the Codex Alimentarius Commission – No. 2: Codes of Hygienic Practice for Dried Fruits.* FAO/WHO, Rome, Italy.

Fellows, P., Hidellage, V. and Judge, E. (1998) *Making Safe Food.* CTA/ITDG Publishing, London.

Fellows, P., Axtell, B. and Dillon, M. (1995) *Quality Assurance for Small-Scale Rural Food Industries.* FAO (ISBN 92 5 103654 3).

Fellows, P. and Axtell, B. (1993) *Appropriate Food Packaging.* TOOL Publications, Amsterdam, The Netherlands (ISBN 90 70857 28 6).

International Trade Centre (1993) *Dehydrated Vegetables: A Survey of Major Markets.* UNCTAD/GATT.

Kindervatter, S. and Range, M. (1986) *Marketing Strategy: Training Activities for Entrepreneurs.* OEF International, Washington, DC (ISBN 0 912917 08 3).

Oti-Boatang, P. and Axtell, B.L. (1996) *Packaging.*, UNIFEM/ITDG Publishing, London.

Robbins, P. (1995) *Tropical Commodities and Their Markets.* Kogan Page Press, London (ISBN 0 7494 1627 0).

Business Development

Fellows, P.J., Franco, E. and Rios, W. (1996) *Starting a Small Food Processing Enterprise.* ITDG Publishing, London (ISBN 1 85339 323 7).

Dickson, D.E.N. (1986) *Improve Your Business.* ILO, Geneva, Switzerland.

Halsall, J.J.H. (19986) *How to Read a Balance Sheet.* ILO, Geneva, Switzerland.

Kindervatter, S. (1991) *Doing a Feasability Study: Training Activities for Starting or Reviewing a Small Business.* UNIFEM, New York.

Technonet Asia (1981) *Entrepreneurs Handbook.* Institute for Small Scale Industries.

Useful contacts

Africa

Botswana – Botswana National Food Technology Research Centre (NFTRC), PO Box 008, Kanye. Fax: 00-267-340713.

Cameroon – Appropriate Technology Transfer Association (ATTA), PO Box 8906, Douala. Tel.: 00-237-47-258800. Fax: 00-237-47-2597.

Ethiopia – Food Research and Development Centre, Ethiopian Food Corporation, PO Box 5688, Addis Ababa.

Ghana – Technology Consultancy Centre (TCC), University of Science and Technology, Kumasi.

Kenya – Department of Food Science and Post Harvest Technology, Jomo Kenyatta University, PO Box 62000, Nairobi. Tel.: 00-254-1-51-52181-4. Fax: 00-254-1-51-52164/52255. e-mail: fsptiku@arcc.or.ke

Kenya – Approtech, PO Box 10973, Nairobi.

Kenya – ITDG, PO Box 39493, Nairobi. Tel.: 00-254-719313/715293/719413. Fax: 00-254-710083.
e-mail: admin@itkenya

Malawi – Malawi Enterprise Development Unit (MEDI), PMB 2, Mpnela.

Nigeria – Department of Food Science & Technology, Federal University of Technology, PMB 1526, Owerri.

Nigeria – International Institute of Tropical Agriculture (IITA), PMB 5320, Ibadan.

Republic of South Africa – Council for Scientific and Industrial Research (CSIR), PO Box 395, Pretoria 0001.

Sudan – ITDG Sudan, PO 4172, Khartoum Central, 1114, Sudan. Tel.: 00-249-11-444-260419. Fax: 00-249-11-472002. e-mail: itsd@sudanmail.net

Tanzania – Small Industries Development Organisation (SIDO), PO 2476, Dar-es-Salaam. Tel.: 00-255-276914/4. Fax: 00-255-21011. e-mail: sido@intafrica.com.

Uganda – Department of Food Science & Technology, Makarere University, PO Box 7062, Kampala.

Uganda – Midway Centre, PO Box 22 3505, Kampala. Tel./Fax: 00-256-41223505. e-mail: midway@imul.com

Zambia – Small Scale Industries Association, PO Box 37156, Lusaka. Tel.: 00-260-1-288434/252150.

Zimbabwe – ENDA, PO Box 3492, Harare.

Zimbabwe – ITDG Zimbabwe, PO Box 1744, Harare. Tel.: 00-263-4-91-403896. Fax: 00-263-4-669773. e-mail: itdg@internet.co.zw

Caribbean and Pacific

Antigua – Chemistry & Food Technology Division, Ministry of Agriculture, Fisheries and Lands, Dunbars, Friars Hill Road, St. John's. Tel.: 001-268-462-4502. Fax: 001-268-462-6281. e-mail: moa@candw.ag

Barbados – Barbados Casse Consultants Ltd., #3 Holders Plantation, St. James. Tel.: 001-246-432-5880. Fax: 001-246-432-5882. e-mail: casse@caribsurf.com

Barbados – Ministry of Agriculture and Rural Development, Graeme Hall, Christ Church. Tel.: 001-246-428-4150; 001-246-428-0061. Fax: 001-246-428-0152.

Barbados – Caribbean Development Bank, Technology Services, PO Box 408, Wildey, St. Michael. Tel.: 001-246-431-1690. Fax: 001-246-426-7269. e-mail: Harvey@caribank.org

Belize – Projects Unit, Ministry of Agriculture, Fisheries and Cooperatives, West Block, Belmoplan. Tel.: 00-501-8-22241/42. Fax: 00-501-8-222409. e-mail: mafpeau@btl.net

Dominica – Produce Chemist's Laboratory, Ministry of Agriculture and the Environment, Botanical Gardens, Roseau. Tel.: 001-767-448-2401 (Ext. 3426). Fax: 001-767-448-7999. e-mail: parbel@hotmail.com

Grenada – Produce Chemist's Laboratory, Tanteen, St. George's. Tel.: 00-473-440-3273/0105. Fax: 00-473-440-3273. e-mail: guimacel@caribsurf.com

Guyana – Agricultural Projects Unit, Ministry of Agriculture, Regent Street/Vissengen Road, Georgetown. Tel.: 00-592-2-60393. Fax: 00-592-2-75357. e-mail: nisa@sdnp.org.gy

Haiti – Programme Transformation Unit, CRDA/MARNDR, Route Nationale #1, Damien, Port-au-Prince. Tel.: 00-509-223-8215.

Jamaica – Food Technology Institute, Scientific Research Council, Hope Gardens, PO Box 350, Kingston 6. Tel.: 001-876-977-9316. Fax: 001-876-977-2194. e-mail: fithead@cwjamaica.com

Jamaica – Applied and Food Chemistry, Department of Chemistry, University of the West Indies, Mona, Kingston 7. Tel.: 001-876-927-1910. Fax: 001-876-977-1835. e-mail: dminott@uwimona.edu.jm

Jamaica – Jamaica Bureau of Standards, 6 Winchester Road, PO Box 113, Kingston 10. Tel.: 001-876-926-3140/45; 001-876-929-4247. Fax: 001-876-926-929-4736.

Papua New Guinea – Appropriate Technology Development Institute, Papua New Guinea University of Technology, PO Box 793, Lae.

Saint Kitts – St. Kitts-Nevis Multipurpose Laboratory, Department of Agriculture, PO Box 39, Basterre. Tel.: 001-869-465-5279. Fax: 001-869-465-3852.
e-mail: mplbos@caribsurf.com

Saint Lucia – Produce Chemist's Laboratory, Research and Development Division, Ministry of Agriculture, Fisheries and Forestry, Block A, Waterfront, Castries. Tel.: 001-758-450-2375; 001-758-450-3206. Fax: 001-758-450-1185.
e-mail: research@slumaffe.org

Saint Vincent & the Grenadines – Home/Farm Management Unit, Ministry of Agriculture and Labour, Richmond Hill, Kingstown. Tel.: 001-809-457-2934. Fax: 001-809-457-1688.

Suriname – Agricultural Experimental Station, Ministry of Agriculture, Animal Husbandry and Fisheries, Hetitia Vriesdelaan #10, Paramaribo. Tel.: 00-597-472442. Fax: 00-597-470301. e-mail: seedunit@sr.net

Trinidad and Tobago – Caribbean Industrial Research Institute (CARIRI), Tunapuna Post Office, Trinidad. Tel.: 001-868-662-7161. Fax: 001-868-662-7177.
e-mail: cariri@trinidad.net

Trinidad and Tobago – Food Science & Technology Unit, Department of Chemical Engineering, Faculty of Engineering, University of The West Indies (UWI), St. Augustine. Tel.: 001-868-645-3232; 001-868-645-3237. Fax: 001-868- 662-4414. e-mail: gbaccust@eng.uwi.tt

Trinidad and Tobago – Department of Food Production, Faculty of Agriculture and Natural Sciences, University of The West Indies (UWI), St. Augustine. Tel.: 001-868-662-2002 (Ext. 2110/2090); 001-868-645-3232/4 (Ext. 2110/2090). Fax: 001-868-663-9686. e-mail: istre@carib-link.net

Europe

Germany – Innotech, Hohenheim University, Institute for Agricultural Engineering in the Tropics, Brandenburger Strasse 2, D-71229, Leonberg. e-mail: innotech.ing@t-online.de

France – Groupe de Recherche et d'Echanges Technologiques (GRET), 213 Rue Lafayette, Paris, 75010.

Italy – United Nations Food and Agricultural Organization (FAO), Via delle Terme di Caracalla, 00100, Rome.

Netherlands – KIT Royal Tropical Institute, Mauritskade 63, 1092 AD, Amsterdam.

United Kingdom – Midway Technology, St Ostwalds Barn, Clifford, Hay on Wye, HR3 5HB.

United Kingdom – Natural Resources Institute (NRI), Central Avenue, Chatham, Kent, ME4 4TB.

United Kingdom – Commonwealth Secretariat, Malborough House, Pall Mall, London, SW1 5HX.

United Kingdom – Intermediate Technology Development Group (ITDG), Bourton Hall, Bourton on Dunsmore, Rugby, CV23 9QZ. Tel.: 00-44-1926-634400. Fax: 00-44-1926-634401. e-mail: infoserve@itdg.org.uk

United Kingdom – International Federation for Alternative Trade (IFAT), 30 Murdock Road, Bicester, Oxon, OX26 4RF. Tel.: 00-44-1869-249819. Fax: 00-44-1869-246381. e-mail: info@ifat.org.uk

Index

118

www.ingramcontent.com/pod-product-compliance
Lightning Source LLC
Jackson TN
JSHW062202130125
77033JS00018B/602